Laboratory Manual

STUDENT EDITION

Glencoe Science

Earth Science
Geology, the Environment, and the Universe

NATIONAL GEOGRAPHIC

McGraw Hill Glencoe

New York, New York   Columbus, Ohio   Chicago, Illinois   Woodland Hills, California

The McGraw-Hill Companies

 Glencoe

Send all inquiries to:
Glencoe/McGraw-Hill
8787 Orion Place
Columbus, OH 43240-4027

ISBN: 978-0-07-879197-0
MHID: 0-07-879197-9

Printed in the United States of America.

1 2 3 4 5 6 7 8 9 10   009   11 10 09 08 07

Contents

Contents, *continued*

Contents, *continued*

Contents, *continued*

How to Use This Laboratory Manual

Working in the laboratory throughout the course of the year can be an enjoyable part of your Earth science experience. ***Earth Science: Geology, the Environment, and the Universe,*** *Laboratory Manual* is a tool for making your laboratory work both worthwhile and fun. The laboratory activities are designed to fulfill the following purposes:

• to stimulate your interest in science in general and especially in Earth science

• to reinforce important concepts studied in your textbook

• to allow you to verify some of the scientific information learned during your Earth science course

• to allow you to discover for yourself Earth science concepts and ideas not necessarily covered in class or in the textbook readings

• to acquaint you with a variety of modern tools and techniques used by today's Earth scientists

Most importantly, the laboratory activities will give you firsthand experience in how a scientist works.

The activities in this manual are of three types: Investigation, Mapping, or Design Your Own. In an Investigation activity, you will be presented with a problem. Then, through use of scientific methods, you will seek answers. Your conclusions will be based on your observations alone or on those made by the entire class, recorded experimental data, and your interpretation of what the data and observations mean. In a Mapping activity, you will use existing maps or create your own to help you solve or understand various problems in Earth science. Some of the labs are called Design Your Own, which are similar to the Design Your Own labs in your textbook. In Design Your Own labs, you will design your own experiments to find answers to problems.

In addition to the activities, this laboratory manual has several other features—a description of how to write a lab report, diagrams of laboratory equipment, and information on safety that includes first aid and a safety contract. Read the section on safety now. Safety in the laboratory is your responsibility. Working in the laboratory can be a safe and fun learning experience. By using ***Earth Science: Geology, the Environment, and the Universe,*** *Laboratory Manual*, you will find Earth science both understandable and exciting. Have a good year!

Writing a Laboratory Report

When scientists perform experiments, they make observations, collect and analyze data, and formulate generalizations about the data. When you work in the laboratory, you should record all your data in a laboratory report. An analysis of data is easier if all data are recorded in an organized, logical manner. Tables and graphs are often used for this purpose.

A written laboratory report should include all of the following elements.

TITLE: The title should clearly describe the topic of the report.

HYPOTHESIS: Write a statement to express your expectations of the results and as an answer to the problem statement.

MATERIALS: List all laboratory equipment and other materials needed to perform the experiment.

PROCEDURE: Describe each step of the procedure so that someone else could perform the experiment following your directions.

RESULTS: Include in your report all data, tables, graphs, and sketches used to arrive at your conclusions.

CONCLUSIONS: Record your conclusions in a paragraph at the end of your report. Your conclusions should be an analysis of your collected data.

Read the following description of an experiment. Then answer the questions.

Mass movements of Earth materials can cause damage to property and lives. The movements are influenced by several factors, such as gravity, a material's resistance to flow, and water. A geologist experimented with different Earth materials to determine how water impacted their soil, and placed the materials separately on three boards. The boards were tilted at an angle of 15°. Beginning with the clay, the geologist carefully poured one liter of water down the board and measured the rate of movement of the clay. Using the same amount of water and the same rate of water flow, she repeated the experiment on the gravel and grass-covered soil. She conducted three trials on each material, recording her data in a table. Lastly, she plotted rates of movement on a graph.

1. What was the purpose of this experiment?

_____

2. What materials were needed for this experiment?

_____

3. Write a step-by-step procedure for this experiment.

_____

_____

_____

_____

_____

_____

Writing a Laboratory Report, *continued*

4. Table 1 shows the data collected in this experiment. Based on these data, state a conclusion for this experiment.

_____

_____

_____

_____

Table 1

Material on Board	Rate of Movement of Material
Clay-covered board	
Gravel-covered board	
Soil-covered board	

5. Plot the data in Table 1 on a bar graph. Show rate of movement on the vertical axis. Use a different colored pencil for each material plotted.

Laboratory Equipment

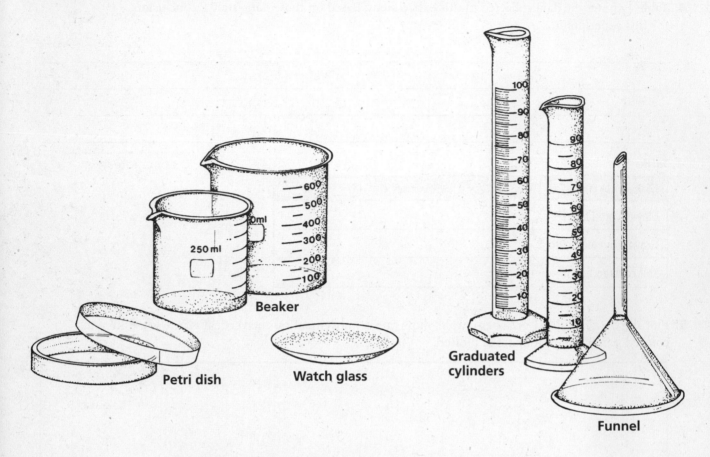

250 ml

Beaker

Petri dish

Watch glass

Graduated cylinders

Funnel

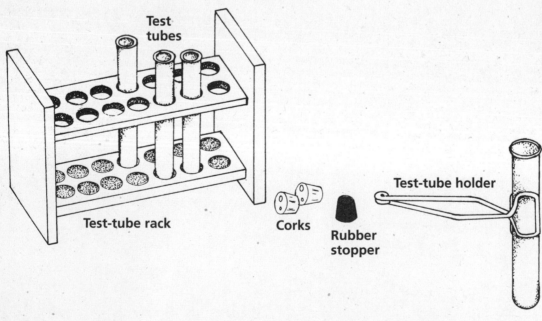

Test tubes

Test-tube holder

Test-tube rack

Corks

Rubber stopper

Laboratory Equipment, *continued*

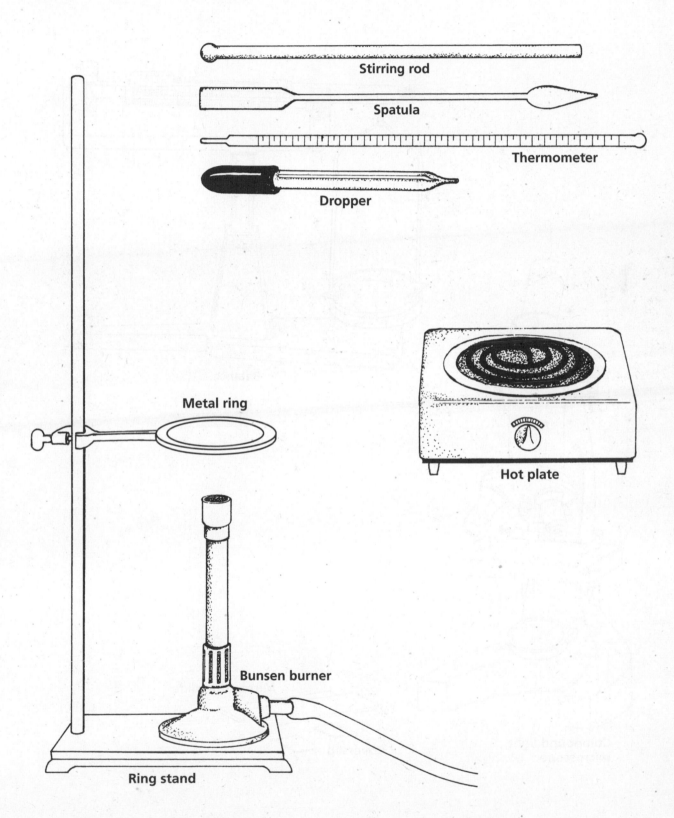

Stirring rod

Spatula

Thermometer

Dropper

Metal ring

Hot plate

Bunsen burner

Ring stand

Laboratory Equipment, *continued*

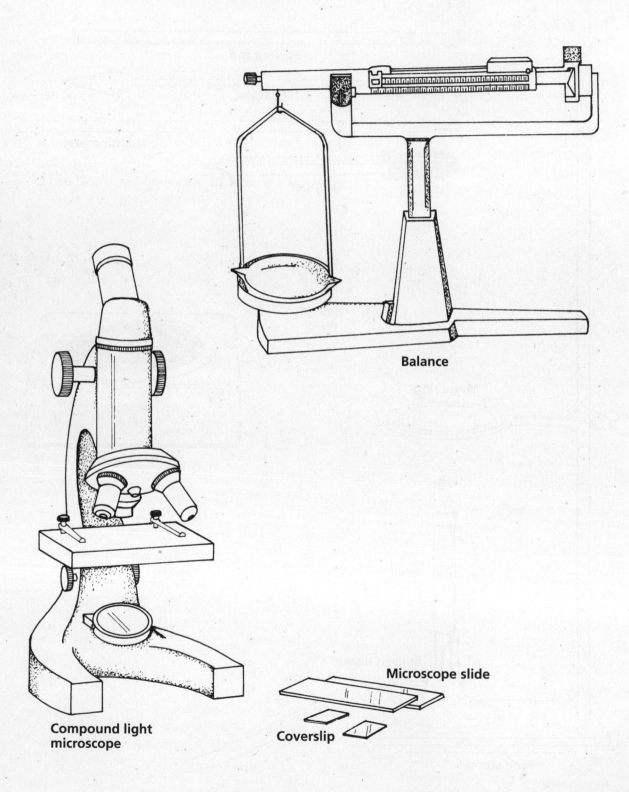

Balance

Microscope slide

Compound light microscope

Coverslip

Safety in the Laboratory

1. Always obtain your teacher's permission to begin a lab.

2. Study the procedure. If you have questions, ask your teacher. Be sure you understand all safety symbols shown.

3. Use the safety equipment provided for you. Goggles and a laboratory apron should be worn when any lab calls for using chemicals.

4. When you are heating a test tube, always slant it so the mouth points away from you and others.

5. Never eat or drink in the lab. Never inhale chemicals. Do not taste any substance or draw any material into your mouth.

6. If you spill any chemical, immediately wash it off with water. Report the spill immediately to your teacher.

7. Know the location and proper use of the fire extinguisher, safety shower, fire blanket, first aid kit, and fire alarm.

8. Keep all materials away from open flames. Tie back long hair and loose clothing.

9. If a fire should break out in the lab, or if your clothing should catch fire, smother it with the fire blanket or a coat, or get under a safety shower. **NEVER RUN.**

10. Report any accident or injury, no matter how small, to your teacher.

Follow these procedures as you clean up your work area.

1. Turn off the water and gas. Disconnect electrical devices.

2. Return materials to their places.

3. Dispose of chemicals and other materials as directed by your teacher. Place broken glass and solid substances in the proper containers. Never discard materials in the sink.

4. Clean your work area.

5. Wash your hands thoroughly after working in the laboratory.

First Aid in the Laboratory

Injury	Safe response
Burns	Apply cold water. Call your teacher immediately.
Cuts and bruises	Stop any bleeding by applying direct pressure. Cover cuts with a clean dressing. Apply cold compresses to bruises. Call your teacher immediately.
Fainting	Leave the person lying down. Loosen any tight clothing and keep crowds away. Call your teacher immediately.
Foreign matter in eye	Flush with plenty of water. Use an eyewash bottle or fountain.
Poisoning	Note the suspected poisoning agent and call your teacher immediately.
Any spills on skin	Flush with large amounts of water or use safety shower. Call your teacher immediately.

Safety Contract

I, _____ , have read and understand the safety rules and first aid information listed above. I recognize my responsibility and pledge to observe all safety rules in the science classroom at all times.

_____   _____

signature   *date*

The *Earth Science: Geology, the Environment, and the Universe* program uses safety symbols to alert you and your students to possible laboratory dangers. These symbols are provided in the student text in Appendix B and are explained below. Be sure your students understand each symbol before they begin an activity that displays a symbol.

SAFETY SYMBOLS	HAZARD	EXAMPLES	PRECAUTION	REMEDY
DISPOSAL	Special disposal procedures need to be followed.	certain chemicals, living organisms	Do not dispose of these materials in the sink or trash can.	Dispose of wastes as directed by your teacher.
BIOLOGICAL	Organisms or other biological materials that might be harmful to humans	bacteria, fungi, blood, unpreserved tissues, plant materials	Avoid skin contact with these materials. Wear mask or gloves.	Notify your teacher if you suspect contact with material. Wash hands thoroughly.
EXTREME TEMPERATURE	Objects that can burn skin by being too cold or too hot	boiling liquids, hot plates, dry ice, liquid nitrogen	Use proper protection when handling.	Go to your teacher for first aid.
SHARP OBJECT	Use of tools or glassware that can easily puncture or slice skin	razor blades, pins, scalpels, pointed tools, dissecting probes, broken glass	Practice common-sense behavior and follow guidelines for use of the tool.	Go to your teacher for first aid.
FUME	Possible danger to respiratory tract from fumes	ammonia, acetone, nail polish remover, heated sulfur, moth balls	Make sure there is good ventilation. Never smell fumes directly. Wear a mask.	Leave foul area and notify your teacher immediately.
ELECTRICAL	Possible danger from electrical shock or burn	improper grounding, liquid spills, short circuits, exposed wires	Double-check setup with teacher. Check condition of wires and apparatus.	Do not attempt to fix electrical problems. Notify your teacher immediately.
IRRITANT	Substances that can irritate the skin or mucous membranes of the respiratory tract	pollen, moth balls, steel wool, fiberglass, potassium permanganate	Wear dust mask and gloves. Practice extra care when handling these materials.	Go to your teacher for first aid.
CHEMICAL	Chemicals that can react with and destroy tissue and other materials	bleaches such as hydrogen peroxide; acids such as sulfuric acid, hydrochloric acid; bases such as ammonia, sodium hydroxide	Wear goggles, gloves, and an apron.	Immediately flush the affected area with water and notify your teacher.
TOXIC	Substance may be poisonous if touched, inhaled, or swallowed.	mercury, many metal compounds, iodine, poinsettia plant parts	Follow your teacher's instructions.	Always wash hands thoroughly after use. Go to your teacher for first aid.
FLAMMABLE	Open flame may ignite flammable chemicals, loose clothing, or hair.	alcohol, kerosene, potassium permanganate, hair, clothing	Avoid open flames and heat when using flammable chemicals.	Notify your teacher immediately. Use fire safety equipment if applicable.
OPEN FLAME	Open flame in use, may cause fire.	hair, clothing, paper, synthetic materials	Tie back hair and loose clothing. Follow teacher's instructions on lighting and extinguishing flames.	Always wash hands thoroughly after use. Go to your teacher for first aid.

 Eye Safety Proper eye protection should be worn at all times by anyone performing or observing science activities.

 Clothing Protection This symbol appears when substances could stain or burn clothing.

 Animal Safety This symbol appears when safety of animals and students must be ensured.

 Radioactivity This symbol appears when radioactive materials are used.

 Handwashing After the lab, wash hands with soap and water before removing goggles

Lab Safety Form

␣e: _____

␣e: _____

type (circle one) : Launch Lab   MiniLab   GeoLab

Title: _____

␣d carefully the entire lab and then answer the following questions. Your teacher must initial ␣form before you begin the lab.

What is the purpose of the investigation?

_____

_____

_____

Will you be working with a partner or on a team? _____

Is this a design-your-own procedure? Circle: Yes   No

Describe the safety procedures and additional warnings that you must follow as you perform this investigation.

_____

_____

_____

Are there any steps in the procedure or lab safety symbols that you do not understand? Explain.

_____

_____

_____

LAB ◀ **1.1** ▶ **INVESTIGATION**

Observing and Analyzing Stream Flow

One way that scientists learn about the world is by modeling natural phenomena and observing and describing what happens. For example, Earth scientists can model how flowing water moves soil by using a stream table, which is a large, shallow pan that is propped at an angle and partly filled with sand or other material (Figure 1). Water is allowed to flow from the higher end of the stream table down through the sand to the lower end of the table. By observing the water's path and its effects on the sand and by altering the rate of water flow, scientists can learn a great deal about stream development.

PREPARATION

PROBLEM
What conclusions can you draw about some of the processes represented in a stream table?

OBJECTIVES
- **Observe** a model of natural phenomena.
- **Communicate** observations clearly and accurately.
- **Choose** criteria to classify observed phenomena.

MATERIALS
pen or pencil
Figures 2–7 in lab manual

PROCEDURE

1. Examine Figures 2–7. They are diagrams of stream development as modeled on a stream table. The table shown is about 4 meters long and 1 meter wide. Before the modeling began, a mass of sand sloped gently and smoothly from the back of the table to about the halfway point on the table. The figures show the water's path and the transport and deposition of sand that occurred as water trickled from a spigot for 2 days.

2. Write a detailed description in the Analyze section on page 3 of what you observe in the figures.

Figure 1

INVESTIGATION

DATA AND OBSERVATIONS

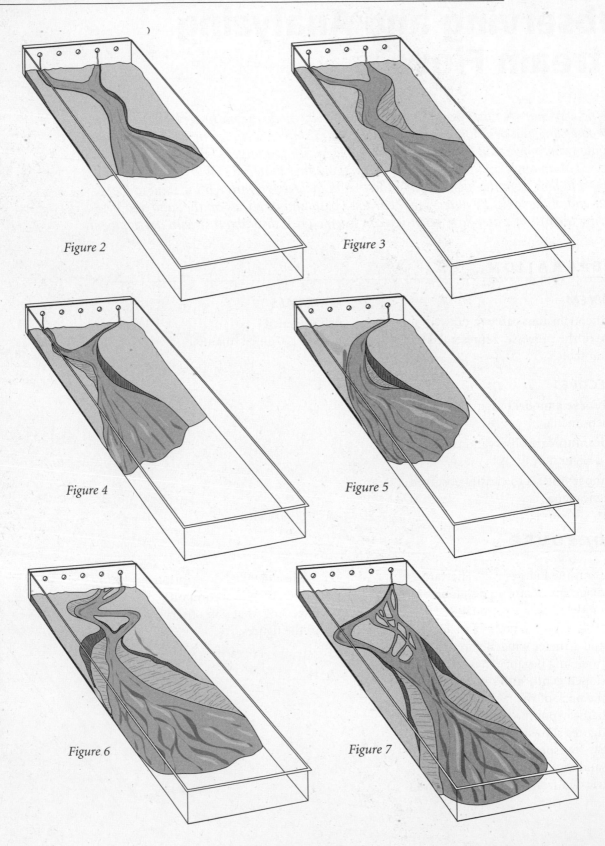

Figure 2

Figure 3

Figure 4

Figure 5

Figure 6

Figure 7

LAB 1.1 INVESTIGATION

ANALYZE

Figure 2

_____
_____
_____
_____

Figure 3

_____
_____
_____
_____

Figure 4

_____
_____
_____
_____

Figure 5

_____
_____
_____
_____

Figure 6

_____
_____
_____
_____

Figure 7

_____
_____
_____
_____

1. Based on your descriptions, make up a classification scheme for the figures. You may decide to classify them all in the same way. Or you may group individual figures into two or more classifications. Describe your classification system.

_____
_____
_____
_____
_____
_____

LAB 1.1   **INVESTIGATION**

CONCLUDE AND APPLY

1. Describe the process of stream development based on your observations of Figures 2–7.

_____

_____

_____

_____

_____

_____

_____

2. What modifications could you make to the stream table to gain further insight into stream development?

_____

_____

_____

_____

_____

LAB ◆ **1.2**   **DESIGN YOUR OWN**

Formulating a Hypothesis

A *scientific method is a planned, organized approach to solving a problem. While the steps taken to solve the problem can vary, the first step involved in scientific problem solving is usually determining what it is you want to know. Often scientific problem solving involves researching the problem. Once the problem is defined and research is complete, a hypothesis, or suggested explanation for an observation, is made. Often the hypothesis is stated in the form of a question that can be answered by the results of a test or an experiment.*

PREPARATION

PROBLEM

How can you create a phenomenon on which a hypothesis can be based and formulate a hypothesis to explain a phenomenon created by others?

OBJECTIVES

- **Design** a hidden phenomenon in a box.
- **Use** various observational methods to **examine** the unknown phenomenon.
- **Write** a hypothesis to **explain** or **describe** the phenomenon.

POSSIBLE MATERIALS

box
small objects
modeling clay
string
bubble wrap
newspaper
tape
magnet
balance or scale
measuring tape
scissors

SAFETY PRECAUTIONS

Be careful when using scissors or handling other sharp objects. Any objects with sharp points or edges could puncture your skin.

LAB 1.2   **DESIGN YOUR OWN**

PLAN THE EXPERIMENT

With your group, discuss how you can use the materials to create a hidden phenomenon. How might the phenomenon be observed by another person? Visually? By touch? By weighing or measuring the box? By smelling it? By shaking or otherwise moving the box? Choose items accordingly, and construct a hidden phenomenon in a box. Describe the contents and arrangement of the box your group designs.

What kinds of observations might you make about an unknown phenomenon to form a hypothesis about what it is? Use Table 1 to organize your ideas. Write a procedure for investigating an unknown phenomenon. After you design the phenomenon in a box, exchange with another group and investigate its box, following your plan.

DATA AND OBSERVATIONS

Test or Observation	Result or Description	Possible Conclusion

LAB ◁ **1.2** ▷   **DESIGN YOUR OWN**

ANALYZE

1. What was the most significant observation you made about the box containing the unknown contents examined by your group?

_____

_____

_____

_____

2. How did you combine observations to draw conclusions about the contents of the box?

_____

_____

_____

_____

_____

_____

LAB ◁ **1.2** ▷   **DESIGN YOUR OWN**

CONCLUDE AND APPLY

1. Based on your observations of the unknown phenomenon and your analysis of these observations, state your hypothesis about the contents and structure of the box.

_____

_____

_____

_____

2. While formulating a hypothesis about the unknown contents of the box you examined, did you apply anything you learned while creating your group's box ? If so, what?

_____

_____

_____

_____

3. How well did your hypothesis match the actual contents of the box examined by your group? What could you have done to get more information?

_____

_____

_____

_____

_____

4. How is this activity similar to actual methods that scientists use to explain unknown phenomena?

_____

_____

_____

_____

_____

_____

LAB 2.1   MAPPING

Use with
Section 2.2

Modeling Topographic Maps

Maps are important tools in studies of the physical characteristics of Earth. Spatial relationships and changes in the shapes and sizes of landforms throughout geologic time can be illustrated using a variety of mapping techniques. Topographic maps show changes in elevation on Earth's surface through the use of contour lines, which connect points of equal elevation above sea level. These contour lines provide useful information about the gradient, shape, and height of Earth's landforms.

PREPARATION

PROBLEM

How can you develop a model of a topographic map and use a topographic map to interpret the shape of a landform?

OBJECTIVES

- **Construct** a model of a mountain with a minimum of two different elevations.
- **Use** contour lines on the model to represent changes in elevation.
- **Model** a topographic map by transferring the contour lines to a flat surface.
- **Interpret** a map constructed by another student to **identify** the appropriate model mountain.

MATERIALS

5-gallon aquarium tank
modeling clay
waterproof black marker
metric ruler or meterstick
masking tape
grease pencil
turkey baster
transparency paper
black yarn
water

SAFETY PRECAUTIONS

- Wear an apron when working with clay to help prevent stains on clothing.
- Wear safety goggles during the lab procedure.
- Wipe up any spills immediately.

LAB ◄ **2.1** ═══════════════════════════════════ **MAPPING**

PROCEDURE

1. Use the grease pencil to draw a 5-cm line on each of the four sides of the tank at 2-cm intervals, beginning with 2 cm from the bottom of the tank. You will have a series of marks 2 cm apart from the bottom to the top of the tank on each side. These marks are for the purpose of measuring water level.

2. Orient one side of the tank to the north and mark that side "North."

3. Construct a clay model of a mountain. The model should fit into the tank with the mountain peak no higher than the top of the tank. If you wish to make it more interesting, the mountain can have more than one peak. Name your mountain.

4. When the model has been placed in the tank, tie one strand of black yarn around the entire base of a mountain to simulate a contour line.

5. Slowly add water to the tank to the level of the first 2-cm line. At the surface level of the water, tie another strand of black yarn around the model.

6. Continue adding water and tying black yarn around the model at each 2-cm level until the water has reached the top of the mountain.

7. With the turkey baster, carefully remove all of the water from the tank. Place transparency paper over the top of the tank, centering it over the mountain, and tape it to the sides of the tank.

8. Place the tank on a low table. Look down at the top of the tank and trace the contour lines from the mountain onto the transparency paper. Label the appropriate end of the map "North." Identify your map as instructed by your teacher and place it on a flat surface to dry. IMPORTANT: At this point, do not put the name of your mountain on the map.

DATA AND OBSERVATIONS

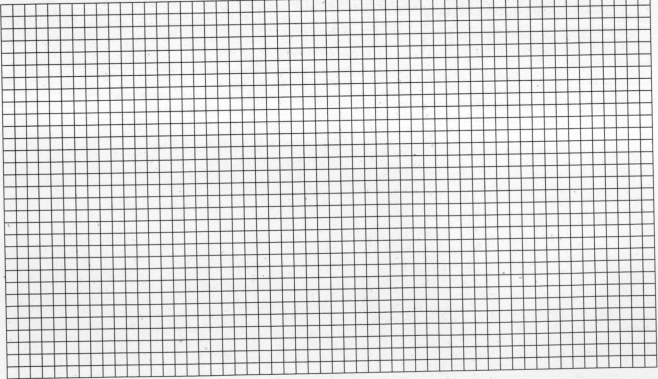

ANALYZE

1. What is the height of your mountain?

_____

2. What is the contour interval of your mountain?

_____

3. Compare the parts of your map that represent the steep gradients with those parts that represent gentler slopes or flat areas. How are the configurations alike or different?

_____

_____

_____

CONCLUDE AND APPLY

1. Exchange maps with another group. You will use this map to identify the group's mountain among the collection of models in your classroom. Observe the map carefully and analyze the shapes and placement of the contour lines. Infer what the mountain might look like. Sketch the mountain in the space below.

LAB ⟨ 2.1 ⟩   **MAPPING**

CONCLUDE AND APPLY, *continued*

2. Orient the map and your sketch to the north. Examine the models of mountains on display in your classroom. Identify the model that is represented by the map.

_____

3. Did you select the correct model?

_____

If not, make a new sketch of the mapped mountain, using the method illustrated in the figure below. On the graph paper in the Data and Observations section, use the transect line method to convert the contour map into a profile. Draw a straight line horizontally across the middle of the contour map. At the points where the contour lines intersect the straight horizontal line (the transect), draw perpendicular lines downward to create the profile. See the figure below. Use this profile to identify the model of the mountain. HINTS: What are the main features indicated by the map? Where are the lines closest together and farthest apart? What do these lines tell you about the shape of the landform?

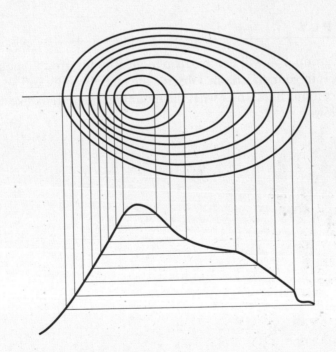

**Use with
Section 2.2**

Interpreting Political and Landform Maps

Among the many uses of maps are those of scientific study and the recording of political, geographic, and geological data for commercial, travel, and recreational purposes. A map is appropriate for a particular purpose only if it contains information in a form consistent with the use to which it will be put. For example, a relief map without political boundaries would be of little use to the tax collector for a specific township, while a road map would not give a hiker or petroleum geologist the full range of information necessary for their pursuits. Although some maps contain more than one type of information, too many different types of data on the same map can be difficult to read. In this investigation, you will be working with political, topographic, and physical maps and analyzing the types of information that each provide.

PREPARATION

PROBLEM
How can different types of maps be prepared and interpreted for various uses?

OBJECTIVES
- **Draw** a political map and a landform map of a single location in or near your school or home.

- **Describe** the strengths and limitations of each map.
- **Compare** the information provided by political, topographic, and physical maps.

MATERIALS
calculator
pencil

PROCEDURE

1. Select an area in or around your school (classroom and hallways area, gym, football field, school neighborhood) or your neighborhood to map.

2. Obtain measurements for drawing the map to scale by pacing the perimeter of the area. A pace is a large step forward. Count the number of paces it takes to travel each side of the area.

3. Decide upon the scale that you will use for your map. Example: Suppose a map is 20 cm on each side. If you measured 43 paces for one side, each centimeter would

represent 2.15 paces (43 divided by 20), and your scale would be 2.15 paces = 1 cm. All of your measurements for the map will then be done in paces and converted to centimeters for the map drawing.

4. Draw the map in the blank space provided. Include a pointer to north, the locations and names of features such as entrances, rooms, corridors, gates, streets, highways, railroads, bodies of water, and town/city boundaries. Label this map "Political."

LAB **2.2** **INVESTIGATION**

PROCEDURE, *continued*

5. Draw a second map of the same area in the blank space provided, using the same perimeter outline from your first map. On this map, draw the physical form of any objects and buildings that exist within the area of the map. Pace off the length and width measurements of these features. From math class, you may know how to determine indirectly the height of buildings, trees, and other tall objects. If not, estimate the heights. Label this map "Physical."

DATA AND OBSERVATIONS

Map 1

Map 2

LAB ◁ **2.2** ▷   **INVESTIGATION**

ANALYZE

1. Compare and contrast the two maps. What similarities and differences do you observe?

_____

_____

_____

_____

2. Which map provides you with the information that you need to travel by car or bicycle? Explain your answer.

_____

_____

_____

_____

3. If you wanted to hike across the area covered by the map in the fastest way possible, which map would be most helpful? Explain your answer.

_____

_____

_____

_____

_____

_____

CONCLUDE AND APPLY

1. Suppose that some landowners allowed hikers to cross their properties but others had high fences and locked gates. What could be added to a map of the properties to make them more useful to hikers?

_____

2. Which of the maps shown below is like your physical map? Which is like your political map? What type of map is Map C?

_____

_____

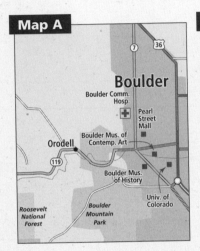

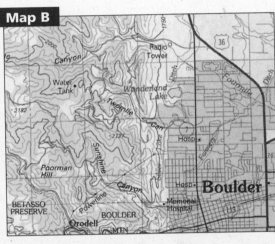

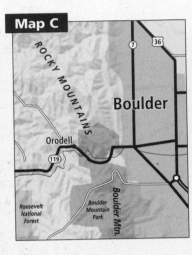

3. If you were in Colorado for the first time and were planning a bicycle trip from Boulder to Orodell, what route would you use if you had only Map A to consult?

_____

_____

_____

4. Would your bicycle route differ if you had Map B or C to consult instead of Map A? Why or why not?

_____

_____

_____

LAB ◀ 3.1 ▶ **DESIGN YOUR OWN**

Use with
Section 3.3

Changes in State

Elements cycle through the lithosphere, hydrosphere, and atmosphere of Earth as different forms of matter change in state. On Earth, most substances are found in three states: solid, liquid, and gaseous. These states reflect the relative amount of thermal energy in a substance. Changes in thermal energy change the speed of molecular motion. For example, when a solid absorbs enough thermal energy from the environment, it melts into a liquid. When a liquid loses enough thermal energy to the environment, it condenses to a solid. Changes in states of matter are important processes on Earth.

PREPARATION

PROBLEM

How can the melting and boiling points of water (a liquid at room temperature) be demonstrated and compared with those of carnauba wax (a solid at room temperature)? Which states of matter will be observed in this demonstration?

OBJECTIVES

- **Describe** methods of measuring the boiling and freezing points of liquids.
- **Plan** and **carry out** a demonstration of changes in state of liquids and solids.
- **Generalize** your results to the scale of the hydrologic, lithospheric, and atmospheric systems of Earth.
- **Predict** the variables involved in further investigation of changes in state.

HYPOTHESIS

Think about how to demonstrate the boiling and freezing points of matter. Determine what method and materials you will use. Make a hypothesis about the nature of the variables involved and what scale of measure will be appropriate for your demonstration. What is the best way to communicate your findings? How can you graph your results?

_____

_____

_____

_____

_____

_____

POSSIBLE MATERIALS

hot plate
ring stand
test-tube clamp
beaker
flask
clear plastic tubing or distillation apparatus
test-tube corks or stoppers
water
test tubes
carnauba wax
alcohol-based thermometer
graph paper

SAFETY PRECAUTIONS

- Do not touch the hot plate while it is on.
- Use caution when handling hot glassware. Wear thermal mitts if needed.
- Melted wax and boiling water can cause burns. Handle carefully.
- Avoid using mercury-based thermometers. Mercury is toxic.
- Wear safety goggles and an apron during the lab procedure.

LAB 3.1 DESIGN YOUR OWN

PLAN THE EXPERIMENT

Brainstorm the steps of the demonstration and any
necessary safety precautions. List the materials you
will use. Prepare charts for recording your data. Select
an appropriate style of graph (line, bar, or circle) for
communicating your results. Which variables will
your graph show? What scale of measurement will
you use? What will the labels for the axes be?

_____

_____

_____

_____

_____

_____

_____

DATA AND OBSERVATIONS

DATA CHART

LAB ◆ **3.1**   **DESIGN YOUR OWN**

ANALYZE

1. How might you improve the method and/or materials?

_____

_____

2. At what time intervals did you record the temperatures?

_____

_____

How did this help you to display your data?

_____

_____

3. What did the graphs show about the behavior of water and wax as they absorb and release thermal energy?

_____

_____

_____

CHECK YOUR HYPOTHESIS

Was your **hypothesis** supported by your data? Why or why not?

_____

_____

LAB **3.1**

DESIGN YOUR OWN

CONCLUDE AND APPLY

1. What conclusion have you reached about changes in state of water and wax?

_____

_____

_____

2. On the graphs of the melting and freezing points of wax, draw a red line around the section of the graph that illustrates the absorption of thermal energy by the wax. Draw a blue line around the section of the graph that illustrates the release of thermal energy. At what point do the two graphs cross? What can you conclude from this data?

_____

_____

_____

_____

_____

_____

3. The following table lists changes in state of matter caused by changes in the amount of thermal energy contained in each substance. Fill in the blank sections to describe the dynamics of these changes.

Form of Matter	State of Matter #1	Change in Thermal Energy (absorbed or released)	State of Matter #2
Ice	Solid	Absorbed	Water
Water	Liquid		Steam
Water	Liquid		Ice
Glue stick			Liquid glue
Liquid glue	Liquid		Bonded glue
Gasoline	Liquid		Gaseous gasoline
Stick of butter		Absorbed at room temperature	Soft butter
Stick of butter	Solid	Absorbed when heated	
Melted chocolate	Liquid		Solid chocolate
Kerosene			Gaseous kerosene
Magma			Solid rock

Rates of Chemical Reactions

The change of one or more substances into other substances is called a chemical reaction. Chemical reactions require different lengths of time for completion, depending on the substances and the conditions for the reaction. While many reactions can be over in a fraction of a second, others, such as those associated with changes in Earth systems, can take much longer. Chemists use collision theory to explain the effects of surface area, concentration, and temperature on reactions.

Below is the equation for the chemical reaction you will observe in this lab.

$$Zn\ (solid) + 2\ HCl\ (aqueous) \Rightarrow ZnCl_2\ (aqueous) + H_2\ (gas)$$

PREPARATION

PROBLEM

Investigate the effects of surface area, concentration, and temperature on the reaction between zinc and hydrochloric acid.

OBJECTIVES

- **Observe** and **record** the results of chemical reactions.
- **Use** collision theory to **interpret** reaction data.
- **Illustrate** the dynamics of chemical reactions in the context of collision theory.
- **Describe** the relationship between the rate of chemical reactions and surface area, concentration, and temperature.

MATERIALS

forceps
6 test tubes
test-tube rack
glass marker
5 large pieces of granulated zinc
several small pieces of granulated zinc
HCl solution A (2 M)
HCl solution B (1 M)
alcohol-based thermometer
beaker
water
hot plate

SAFETY PRECAUTIONS

- Use forceps to handle the zinc.
- Point the open end of the test tube away from you while heating it.
- If the HCl splatters, wash the area immediately with cold water.
- Use caution when handling hot glassware. Wear thermal mitts if needed.
- Avoid using mercury-based thermometers. Mercury is toxic.
- Wear safety goggles and an apron during the lab procedure.

LAB 3.2   **INVESTIGATION**

PROCEDURE

1. Put on the safety goggles, gloves, and a lab apron.

2. Put two clean test tubes in the test-tube rack. Label one tube "A" and the other "B." Drop one of the larger pieces of zinc into each tube.

3. Add HCl solution A to one-quarter of the height of tube A and HCl solution B to the same height in tube B. After a minute or two, observe carefully to see how fast the gas in each tube is bubbling. Record the results in the table below.

4. Put two clean test tubes in the test-tube rack. Label the test tubes "L" and "S." Drop one of the larger pieces of zinc in tube L and several small pieces in tube S.

5. Add HCl solution A to one-quarter of the height of each tube. Observe the rate of bubbling in each tube and record the results.

6. Label two clean test tubes "R" and "H." Drop a large piece of zinc in each test tube. Put test tube R in the rack at room temperature, and place test tube H in a beaker half full of water on a hot plate. Add HCl solution A to one-quarter of the height of each tube.

7. Put a thermometer in test tube H and note the temperature. Heat the tube to raise the temperature by 10°C. Observe the rate of bubbling in each tube and record the results in the table. Turn off the hot plate and put test tube H in the rack.

8. Review the results that you have recorded in your table and, for each set of two test tubes, determine which effect caused the faster reaction rate: surface area, concentration, or temperature. Enter the cause in the last column of the table.

DATA AND OBSERVATIONS

Reactants	Bubbles Faster	Bubbles Slower	Cause of Faster Reaction
Tube A: large zinc + dilute HCl			
Tube B: large zinc + 1/2 dilute HCl			
Tube L: large zinc + dilute HCl			
Tube S: small zinc + dilute HCl			
Tube R: zinc + HCl at room temperature			
Tube H: Zinc + heated HCl			

LAB 3.2   **INVESTIGATION**

ANALYZE

1. The diagrams below show how collision theory explains the effect of concentration on reaction rate. Draw additional diagrams illustrating the effect on reaction rate of temperature and surface area.

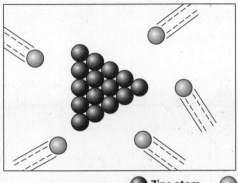

Low concentration of acid

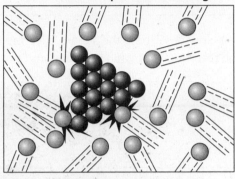

**Concentration higher—
more chance of particles colliding**

● Zinc atom   ○ Hydrogen ion from acid

Low Temperature   **High Temperature**

Small Surface Area   **Large Surface Area**

LAB ◆ **3.2**   **INVESTIGATION**

CONCLUDE AND APPLY

1. What chemical reaction results in the formation of rust?

   _____

2. When metal is painted to prevent the formation of rust, which factor
   (concentration, surface area, or temperature) is being reduced?

   _____

3. Complete the table below describing the factors affecting reaction rate. In the last
   three rows, describe reactions that you observe in and on Earth.

Reaction	Rate Reduced or Increased	Factor Involved
Metal rusting	Reduced by painting	
Acid burning skin	Reduced by rinsing	
Food spoiling		
Apple slices turning brown		
Coffee brewing		
Limestone formations dissolving, producing caves (Hint: Where are limestone caves found?)		

LAB 4.1   **INVESTIGATION**

Growing Crystals

Weak solutions of mineral salts can result when rainwater percolates through rock. When the solution reaches a hole in the rock, such as a fracture, the solution may evaporate and become more concentrated. When the solution becomes saturated, further evaporation deposits some of the salt as crystals. This evaporation is responsible for many common forms of crystal deposition. We can simulate this process by using small amounts of supersaturated solutions.

PREPARATION

PROBLEM
How do crystals form from solutions?

OBJECTIVES
- **Form** crystals by evaporating solutions.
- **Identify** several of the major crystal systems.

MATERIALS
solutions A–D
solutes A–D
4 test tubes
test-tube rack
glass marker
4 small petri dishes
metal spatula
Bunsen burner
lighter
test-tube tongs
pipette
4 microscope slides
table lamp
compound microscope

SAFETY PRECAUTIONS

- Wear safety goggles, gloves, and an apron during the lab procedure. Tie back long hair and do not wear loose clothing.
- Circuits must be protected by a GFI (ground fault interruptor) when electricity is used near a water source.
- Some of the solutions are toxic and may irritate skin; wash your hands at the end of lab. Rinse spills with water.
- Do not inhale any powder or fumes; chemicals can be toxic if inhaled.
- Wear thermal mitts when handling hot objects.
- Follow your teacher's instructions to dispose of used slides and hazardous chemicals.

PROCEDURE

1. Label four test tubes A–D. Half fill each one with the corresponding solution provided by your teacher.

2. Label four small petri dishes A–D. Place a few heaped spatulas of each solute into its corresponding dish.

3. Use tongs to heat test tube A with a Bunsen burner to a temperature that is about midway between room temperature and the boiling point of water (approximately 66°C).

LAB **4.1**   **INVESTIGATION**

PROCEDURE, *continued*

4. Add a spatula of each solute to its corresponding solution. Stir with a clean spatula to dissolve the solute. Repeat this step until nearly all of the solute dissolves.

5. Label four microscope slides A–D. Use a pipette to place a drop of warm supersaturated solution A on slide A. Rinse and dry the pipette. Repeat this step for each solution. Record the time in Table 2.

6. Collect the slides in one place. Bring the bulb of a table lamp within about 15 cm of the collected slides to keep them warm.

7. Observe each slide under the microscope about every 2 minutes as crystals form. Record your observations, including the time at which crystals first appear and how they look. Note how the crystal structure develops.

8. After 15–20 minutes, draw the crystals in the space on the next page.

DATA AND OBSERVATIONS

Table 1

Solution	Chemical	Formula
A	Alum	$AlK(SO_4)_2 \cdot 12H_2O$
B	Rochelle salt	$KNaC_4H_4O_6 \cdot 4H_2O$
C	Copper acetate monohydrate	$Cu(CH_3COO)_2 \cdot H_2O$
D	Calcium copper acetate hexahydrate	$CaCu(CH_3COO)_4 \cdot 6H_2O$

Table 2

Solution	Start Time	Observations
A. Alum		
B. Rochelle salt		
C. Copper acetate monohydrate		
D. Calcium copper acetate hexahydrate		

LAB ◇ **4.1**   **INVESTIGATION**

DATA AND OBSERVATIONS, *continued*

DRAWING OF CRYSTALS

ANALYZE

1. Consider the four examples of crystallization that you observed. Describe the process of evaporative crystallization. Identify trends that apply to all the samples, and summarize those trends.

_____

_____

_____

_____

_____

_____

_____

_____

_____

_____

LAB **4.1**   **INVESTIGATION**

CONCLUDE AND APPLY

1. Summarize in Table 3 the major crystal system and color that results from the evaporation of each solution. Refer to Table 4-1 on page 78 in your textbook.

Table 3

Solution	Crystal System	Color
A. Alum		
B. Rochelle salt		
C. Copper acetate monohydrate		
D. Calcium copper acetate hexahydrate		

2. Compare your model to natural formations of crystals.

_____

_____

_____

_____

_____

_____

_____

LAB **4.2** **DESIGN YOUR OWN**

Use with
Section 4.2

Rockhounding

*R*ocks can tell you a lot about your area's geologic history. Every area has places *that are good for rockhounding, or seeking and collecting rocks. For example, there may be quartz mines where, for a small fee, those searching for rocks can gently hammer away at special seams to obtain delicate crystals. You may be able to find interesting outcrops of rock with minerals embedded in them along timber roads and in gravel bars and creek drainages.*

PREPARATION

PROBLEM

What rocks and minerals are found in your area?

OBJECTIVES

- **Identify** a local spot of geologic interest.
- **Plan** a field trip to collect samples of rocks and minerals.
- **Collect** samples of different rocks and minerals.
- **Identify** rocks and minerals.

HYPOTHESIS

What kinds of rocks and minerals do you expect to find in your area? Write your hypothesis below.

_____

_____

_____

_____

POSSIBLE MATERIALS

resources on local geology
small backpack
chisel-edged hammer
short pry bar
cloth or leather gloves

newspaper
bag for samples
notebook
masking tape
markers
hand lens
field guide to rocks and minerals
food
water
sunscreen
porcelain tile
hardness testing kit
dilute hydrochloric acid
long-sleeved shirt or jacket

SAFETY PRECAUTIONS

- Avoid treacherous terrain.
- Wear safety goggles to protect your eyes from flying chips of rock.
- Use caution when hammering.
- Carefully examine your surroundings. Avoid contact with poisonous plants and animals. Watch out for ticks.
- Wear your cloth or leather gloves during the entire lab procedure.
- Hydrochloric acid is corrosive. Keep it away from your skin and eyes.

LAB 4.2   **DESIGN YOUR OWN**

PLAN THE EXPERIMENT

As a group, research the geology of your area on the Internet, in pamphlets about the local geology, and with geologic maps. Write about two or three locations that are accessible and geologically interesting. As a class, consider all proposals and choose a site to visit, bearing in mind geologic merit and practical considerations, such as time and transport. Once the class has decided where to go, compile a table of a dozen or so rocks and minerals that you might expect to find there. Note a few characteristics of each that will help you with field identifications.

Look at the possible materials. A notebook is essential for writing directions to the site, sketching outcrops, and describing the rocks you collect. Other essentials are water and safety equipment. Plan to collect 10–12 samples. Write a checklist of what you will need to pack at home and what you will need from school.

DATA AND OBSERVATIONS

FIELD IDENTIFICATION TABLE

LAB **4.2**   **DESIGN YOUR OWN**

ANALYZE

1. After the field trip, confirm your field identifications by checking hardness, specific gravity, color, streak, cleavage, luster, and crystal form. Compile a table with an identification number for each sample, your tentative identification, tests performed, and a confirmed identification.

2. As a class, compile a table of everyone's sample information.

3. Exhibit your specimens with informative labels. View your classmates' samples. Take notes on the minerals and rocks that you did not collect yourself.

DATA TABLES

LAB **4.2** **DESIGN YOUR OWN**

CHECK YOUR HYPOTHESIS

Was your **hypothesis** supported by your data? Why or why not?

_____

_____

CONCLUDE AND APPLY

1. What is your most remarkable specimen? What is the most remarkable specimen in the class? Explain your answer.

_____

_____

_____

_____

2. What were the most common types of rocks and minerals at the collection site? Were these the most frequently collected samples?

_____

_____

_____

_____

3. Were the samples what you expected to find? If they were different, explain why.

_____

_____

_____

_____

LAB **5.1**   **INVESTIGATION**

Comparing Lunar Rocks to Earth Rocks

Igneous rocks form when molten rock called magma cools. Igneous rocks are classified by mineral composition and texture. Their mineral composition indicates the nature of the magma. Texture indicates how the magma cooled. Rocks collected from the Moon have characteristics like those of igneous rocks on Earth. These characteristics provide insight about the composition of the Moon and how it was formed.

PREPARATION

PROBLEM

How do lunar rocks compare with Earth rocks?

OBJECTIVES

- **Estimate** mineral percentages in igneous rock samples.
- **Identify** types of igneous rocks.
- **Compare** lunar rocks to Earth rocks.

MATERIALS

4 igneous rocks from Earth
igneous-rock key
pictures of lunar rocks

SAFETY PRECAUTIONS

Use caution when handling rocks; edges can be sharp. Do not wear sandals during the lab procedure.

IGNEOUS-ROCK KEY

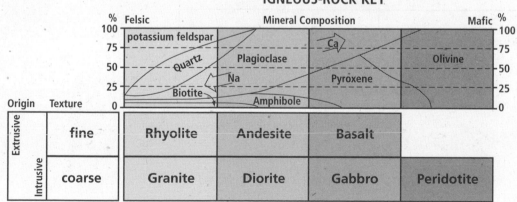

LAB ◁ **5.1** ▷   **INVESTIGATION**

PROCEDURE

1. Observe a rock sample provided by your teacher and determine whether the texture is fine-grained or coarse-grained. Record this data in the table.

2. Estimate and record the percentage of dark minerals. This percentage will allow you to classify the rock as felsic (few dark minerals, light-colored or gray), intermediate (about 50 percent dark minerals, dark gray), or mafic (more than 70 percent dark minerals, very dark or black).

3. Estimate and record the percentage of feldspar. Classify the feldspar as potassium (pink, white, or gray) or plagioclase (white or gray with striations).

4. Estimate and record the percentage of quartz. If the sample has no quartz, it belongs in the gabbro-basalt group. If it has less than 10 percent quartz, it belongs in the diorite-andesite group. If it has 10–40 percent quartz, it belongs in the granite-rhyolite group.

5. Using the igneous-rock key on page 33, identify the unknown igneous-rock sample. Write the correct name of the sample in the table.

6. Repeat steps 1–5 for the other rock samples.

7. Look at pictures of lunar rocks, and use your observations to answer the questions in Analyze and Conclude and Apply.

DATA AND OBSERVATIONS

TABLE

Rock Sample	Texture	Estimated % of Dark Minerals	Felsic, Mafic, or Intermediate	Name of Rock
1				
2				
3				
4				

LAB **5.1** INVESTIGATION

ANALYZE

1. What elements appear to be prevalent in the lunar rocks?

_____

_____

_____

_____

2. Which igneous rocks from Earth are most like the lunar rocks? Explain your answer.

_____

_____

_____

_____

_____

3. What similarities and differences did you notice between lunar rocks and Earth rocks?

_____

_____

_____

_____

_____

LAB **5.1** **INVESTIGATION**

CONCLUDE AND APPLY

1. Scientists theorize that numerous lavaflows have erupted on the surface of the Moon millions of years ago. What evidence proves this?

_____

_____

_____

_____

2. Darkened spots on the Moon's surface are visible from Earth. These dark spots are created by huge fields of mafic rock located deep in the valleys of craters. How do you think these mafic rocks got there? Why are these areas darker than the surface of the Moon?

_____

_____

_____

_____

_____

_____

3. The grain size of lunar rocks is relatively small. What does a fine grain size suggest about how the magma cooled on the surface of the Moon?

_____

_____

_____

_____

LAB ◆ **5.2**   **MAPPING**

**Use with
Section 5.2**

Locating Igneous Rocks on Earth

Intrusive igneous rocks form beneath Earth's crust when magma cools slowly.
Extrusive igneous rocks form on the surface when lava cools very rapidly. Differences
in cooling rate determine the texture of igneous rocks. When magma cools slowly, it
forms large crystals and coarse-grained rocks. When magma cools quickly, it forms
small crystals and fine-grained rocks. When magma cools even more rapidly, such
as in water, it forms extremely small crystals and glassy rocks. At times, two stages
of cooling occur, forming rocks that are porphyritic. These rocks have large crystals
surrounded by small crystals.

    The color of igneous rocks indicates their mineral composition. Dark rocks
usually contain large amounts of iron and magnesium and are called mafic. Light
rocks contain large amounts of feldspar and silicon and are called felsic. Intermediate
rocks are a mixture of light and dark colors.

PREPARATION

PROBLEM
How can you identify where rocks
originate?

OBJECTIVES
- **Classify** igneous rocks based on
  texture and color.
- **Recognize** that the characteristics of
  rocks are linked to their formation
  conditions and origins.
- **Plot** the location of igneous rocks
  on a map.

MATERIALS
9 igneous rocks

SAFETY PRECAUTIONS

Use caution when handling rocks; edges
can be sharp. Do not wear sandals
during the lab procedure.

PROCEDURE

1. Observe nine igneous rock
samples. Using Table 1 and
Figure 1 as guides, record your
observations of color, texture, and
type for each sample in Table 3.

2. Using the coordinates in Table 2,
match each letter with the
appropriate point on the map in
Figure 2. Each labeled point is the
origin of a rock sample that you
have observed.

3. Using your data in Table 3, match
each rock (1–9) with its proper
point on the map. Write the rock
number next to the labeled point
(A–I). It may be possible that a
rock originated in several places.
List each potential point of origin
in Table 3.

LAB **5.2**   **MAPPING**

DATA AND OBSERVATIONS

Table 1

Igneous Rock Type	Common Colors	Minerals
Felsic	white, tan, gray, pink, red	silica, quartz, orthoclase and plagioclase feldspar, some mica and hornblende
Intermediate	gray, green	plagioclase feldspar, hornblende, augite, biotite, amphibole, and pyroxene
Mafic	dark green, dark gray, black	iron, magnesium, plagioclase feldspar, augite, biotite, amphibole, pyroxene, and olivine

Table 2

Location	Latitude (N)	Longitude (W)	Depth (m)
A	21°	102°	−2.5
B	40°	120°	−195
C	27°	73°	−80 below water
D	52°	47°	−120 below water
E	17°	136°	−105 below water
F	38°	79°	−0.5
G	62°	112°	−93
H	28°	142°	−1.5
I	61°	162°	−772

Figure 1

Igneous Rock Textures

glassy   coarse-grained

fine-grained   porphorytic

 **LAB** **5.2**   **MAPPING**

Table 3

Rock	Color(s)	Felsic, Mafic, or Intermediate	Texture (glassy, fine-grained, coarse-grained, or porphyritic)	Intrusive or Extrusive
1				
2				
3				
4				
5				
6				
7				
8				
9				

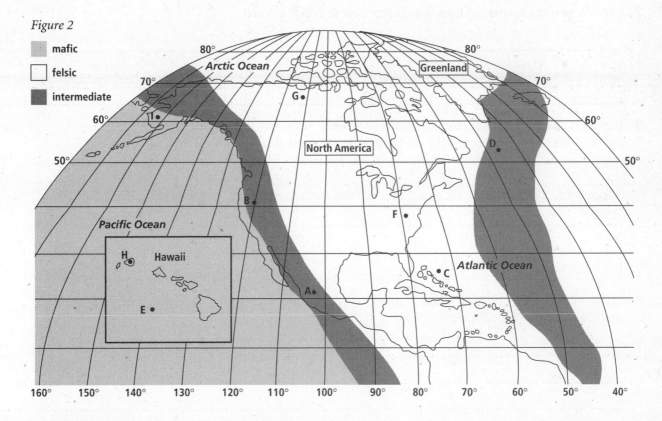

Figure 2

ANALYZE

1. Where on the map is felsic rock found?

LAB **5.2** **MAPPING**

ANALYZE, *continued*

2. Where on the map is mafic rock found?

_____

3. Where on the map is intermediate rock found?

_____

CONCLUDE AND APPLY

1. How did you determine whether the rocks were felsic, mafic, or intermediate?

_____

_____

_____

2. Explain how the texture of a rock indicates how the rock formed.

_____

_____

_____

3. Hawaii is known for its beaches of black sand, while Texas has miles of beaches where the sand is as white as snow. Why do the beaches in these two states have such different types of sand? How do you think this sand was formed?

_____

_____

_____

_____

4. Extrusive rocks may be glassy, fine-grained, or porphyritic. Compare and contrast the conditions under which these rocks form.

_____

_____

_____

_____

_____

_____

LAB ◆ **6.1** ◆ **INVESTIGATION**

Use with Section 6.1

Comparing Chemical Sedimentary Rocks and Modeling Their Formation

There are three main types of sedimentary rock: clastic, organic, and chemical. Clastic sedimentary rocks form when sediments such as sand or clay are cemented together. Organic sedimentary rocks form from the remains of once-living organisms that have been buried and lithified. Chemical sedimentary rocks may form when minerals precipitate out of a solution as a result of evaporation of water. They may also form from chemical reactions of ions in a solution. Common chemical sedimentary rocks include chert, rock salt, rock gypsum, and travertine.

PREPARATION

PROBLEM

How can you distinguish among different types of chemical sedimentary rocks? How do chemical sedimentary rocks form?

OBJECTIVES

- **Differentiate** among several types of chemical sedimentary rocks.
- **Simulate** the formation of chemical sedimentary rocks.

MATERIALS

4 chemical sedimentary rocks
sodium chloride solution
silver nitrate solution
water   500-mL beaker
2 test tubes   hot plate
test-tube rack   dropper
test-tube holder

SAFETY PRECAUTIONS

- Be careful when using the hot plate and handling hot glassware. Wear thermal mitts if needed.
- Silver nitrate stains clothing and skin. Stains on skin cannot be removed. Stains on clothing can be removed with a stain remover. Silver nitrate is highly toxic.
- Wear safety goggles, gloves, and an apron during the lab procedure.

PROCEDURE

1. Fill a beaker three-quarters full of water. Place the beaker on a hot plate and heat it to boiling.

2. Label two test tubes 1 and 2. Use a graduated cylinder to fill each one half full of sodium chloride solution and set the test tubes in a test-tube rack.

PROCEDURE, *continued*

3. Using the test-tube holder, place test tube 1 upright in the boiling water, and hold it in the water without letting it rest on the bottom of the beaker. Heat the test tube until much of the water in the test tube has boiled away and white crystals become visible (about 10 minutes). Record your observations in Table 1.

4. Use a dropper to add two drops of silver nitrate to test tube 2. Watch what happens and record your observations.

5. Now examine the chemical sedimentary rocks provided by your teacher.

DATA AND OBSERVATIONS

Table 1

Test Tube	Observations
1	
2	

Table 2

Rock	Color	Characteristics
Travertine	White or cream	Has dense, closely compacted layers; generally occurs in banded layers
Rock salt	Colorless to white	Has cubic crystals; generally occurs as a mass of intergrown crystals
Rock gypsum	White, gray, brown, red, or green	Is very soft, may have thin layers, generally appears massive, may be crumbly

LAB 6.1 **INVESTIGATION**

ANALYZE

1. Use the information in Table 2 to help identify the chemical sedimentary rock samples provided by your teacher. Record your identifications in Table 3 below.

Table 3

Sample	Rock Name
1	
2	
3	

2. How did the crystals in test tube 1 form?

_____

_____

_____

_____

_____

3. How did you know that a chemical reaction was taking place in test tube 2?

_____

_____

_____

_____

_____

LAB **6.1**   **INVESTIGATION**

CONCLUDE AND APPLY

1. In this lab, you observed the formation of precipitates from a salt solution. The same chemical elements make up common table salt and the mineral halite, or rock salt. What are these elements?

_____

_____

_____

_____

2. What are some of the characteristic features of sedimentary rock?

_____

_____

_____

_____

3. Where do you think you are most likely to find chemical sedimentary rock? Why?

_____

_____

_____

_____

LAB 6.2

MAPPING

Grand Canyon Formations

*T*he Grand Canyon is not only a famous recreational site, but it is the subject of
intense scientific research. The Grand Canyon is composed of many layers of rock,
most of which are sedimentary. These rocks often contain fossils that help scientists
determine how and when the rock formed. If you were to hike through the Grand
Canyon today, you would be exposed to numerous rock layers that represent more
than 2 billion years of geologic history.

PREPARATION

PROBLEM

How can you use a geologic map of the
Grand Canyon to interpret the geologic
history of the area?

OBJECTIVES

- **Interpret** information about rock
  layers in the Grand Canyon.
- **Create** a geologic cross section.
- **Hypothesize** about how rock layers
  formed.

MATERIALS

colored pencils

PROCEDURE

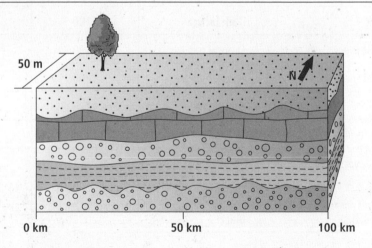

1. Examine the geologic cross section
in the figure above.

2. Using the figure above as a guide,
construct a cross section of the
Grand Canyon rock layers listed in

Table 1. Draw the cross section on
the grid in Data and Observations.

3. Once you have drawn in the rocks,
color each layer to match the colors
in Table 1.

LAB **6.2**   **MAPPING**

PROCEDURE, *continued*

Table 1

Rock Layer	Composition	Color	Age (millions of years)	Thickness (feet) East to West
Kaibab Limestone	Sandy limestone	Grayish white	250	300–500
Toroweap Formation	Sandy limestone	Grayish yellow	255	250–450
Coconino Sandstone	Quartz sand	Cream	260	350–50
Hermit Shale	Shale	Rusty red	265	250–1000
Supai Formation	Shale	Red	285	950–1350
Redwall Limestone	Marine limestone	Red	335	450–700
Temple Butte Limestone	Freshwater limestone	Cream	350	0–450
Muav Limestone	Limestone	Gray	515	400–1000
Bright Angel Shale	Mudstone shale	Greenish brown	530	300–450
Tapeats Sandstone	Sandstone	Dark brown	545	250–150
Great Unconformity	Rock layers eroded or never deposited			
Chuar Group	Sandstone Shale Limestone	Tan Black Green	825–1000	6900
Nankoweap Formation	Sandstone	Gray	1050	6900
Cardenas Basalt	Basalt	Dark brown	1100	980
Dox Sandstone	Sandstone	Orange red	1190	3000
Shinumo Quartzite	Sandstone	Purplish brown	1200	1070–1560
Hakatai Shale	Shale	Orange red	1225	430–830
Bass Formation	Limestone	Grayish	1250	120–340
Early Unconformity	Rock layers eroded or never deposited			
Zoroaster Granite	Granite	Dark gray	1700–1900	?
Vishnu Schist	Mica schist	Black	2000	?

LAB 6.2

MAPPING

DATA AND OBSERVATIONS

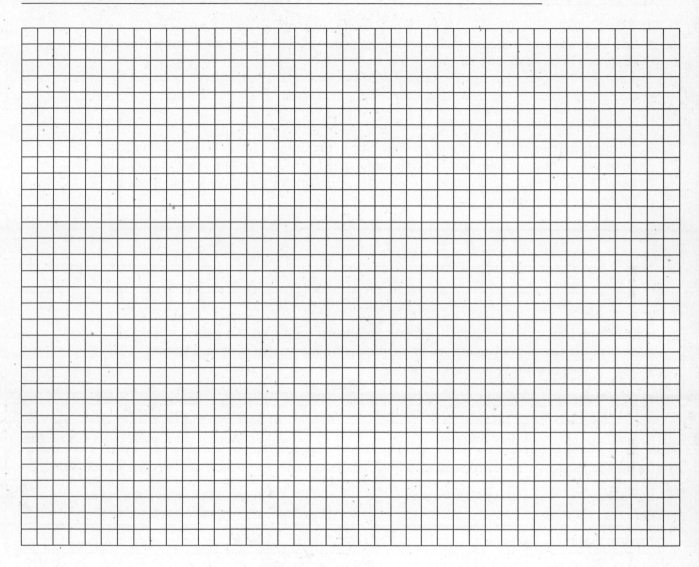

ANALYZE

1. What problems did you have with the vertical scale of the cross section? Will the cross section be the same in a different area? How will it change?

_____

_____

_____

2. Which layers in the Grand Canyon are metamorphic rocks? What factors contributed to their formation?

_____

_____

LAB **6.2**   **MAPPING**

CONCLUDE AND APPLY

1. Over 2 billion years ago, the area that is now to the north of the Grand Canyon in Colorado and Utah was once a mountain range taller and wider than the Rocky Mountains. What happened to those mountains over time?

_____

2. Scientists have found two spans of geologic time in the Grand Canyon for which no rock layers exist. These are called the Great Unconformity (~550–820 million years ago) and the Early Unconformity (~1255–1695 million years ago). How are the unconformities related to changes in depositional environment?

_____

_____

_____

Scientists have been able to hypothesize the environmental conditions that existed when the layers of the Grand Canyon formed by examining the characteristics of each layer. Use Table 2 to answer the following questions.

Table 2

Rock Layer	Composition	Fossils
Redwall Limestone	Marine limestone	Brachiopods, clams, snails, corals, fish, trilobites
Coconino Sandstone	Pure quartz sand, basically a petrified sand dune	No bone fossils; invertebrate tracks and burrows
Hermit Shale	Soft, easily eroded shale	Ferns, conifers, other plants; reptile and amphibian tracks; no bones
Zoroaster Granite	Granite	None

3. Hypothesize about the environment that existed when the Coconino Sandstone formed. Give reasons for your hypothesis.

_____

_____

4. What does the information given about the Hermit Shale suggest about the environment that existed when it formed?

_____

_____

5. Why do you suppose that no fossils are present in Zoroaster Granite?

_____

_____

LAB **7.1**   **INVESTIGATION**

Use with
Section 7.1

Chemical Weathering and Temperature

Chemical weathering takes place when rocks and minerals undergo changes in their composition as a result of a chemical reaction. Significant agents of chemical weathering include water, oxygen, carbon dioxide, and acids. Acids contribute to the chemical weathering of limestone, which produces carbon dioxide gas as the calcium carbonate dissolves. This gas production may be too slow to observe, but the loss of mass can be measured with a balance.

PREPARATION

PROBLEM
What effect does temperature have on the chemical weathering of limestone?

OBJECTIVES
- **Determine** the effect of an acid on limestone.
- **Model** the effect of temperature on the chemical weathering of limestone.
- **Calculate** the relationship between temperature increase and chemical breakdown.

MATERIALS
100 g pea-sized limestone chips
balance
paper towels
glass marker
200-mL beakers (2)
500-mL beakers (2)
400 mL vinegar

100-mL graduated cylinder
ice
aluminum foil
alcohol-based thermometer
hot water
plastic spoons

SAFETY PRECAUTIONS

- Vinegar is a weak acid and can burn sensitive skin. Wear your goggles at all times.
- Avoid using mercury-based thermometers. Mercury is toxic.
- Wash your hands after handling the limestone chips and vinegar.
- Label all solutions.
- Wear an apron during the lab procedure.
- Follow your teacher's suggestions for disposing of lab materials.

PROCEDURE

1. Get about 100 g of limestone chips. Make two piles of chips, on two paper towels, and blot them as dry as possible.

2. Label two 200-mL beakers "C" and "W." Weigh one of the two piles of

chips to the nearest 0.1 g, and put them in beaker C. Record the mass in the table provided. Weigh the other pile of chips and put them in beaker W. Record the mass.

LAB **7.1**

PROCEDURE, *continued*

3. Add about 100 mL of cold vinegar to beaker C and 100 mL of warm vinegar to beaker W. Put about 400 mL of an ice-water mixture in one 500-mL beaker, and about 400 mL of hot water in the other beaker. **CAUTION:** *Direct contact with hot water can burn skin.* Place beaker C in the ice-water mixture, and beaker W in the hot water. Measure and record the temperature in beakers W and C.

4. Look for evidence of a chemical reaction in the two beakers. Record what you observe in the table.

5. Cover both beakers with aluminum foil. After 20–30 minutes, use a plastic spoon to fish out the limestone pieces from each beaker. Place them on separate labeled paper towels and blot them dry.

6. Measure and record the mass of the chips from both beakers.

7. Put the chips back into their respective beakers. Add 100 mL of cold vinegar to the chips in beaker C and 100 mL of warm vinegar to the chips in beaker W. Cover both beakers with aluminum foil.

8. Refrigerate beaker C. Put beaker W in the lab where it can sit overnight.

9. The next day, record the temperature of the vinegar in the two beakers. Pour out the vinegar and rinse away small black pieces and residue. Place the chips on separate labeled paper towels and blot them dry.

10. Measure and record the mass of each pile of chips.

11. Calculate and record the percentage change in mass.

12. Use the figure following the table to create a bar graph of your results.

DATA AND OBSERVATIONS

TABLE

		Beaker C	Beaker W
	Mass before weathering		
	Mass after weathering		
Day 1	Temperature of vinegar		
	Change in mass		
	Percentage change in mass		
	Mass before weathering		
	Mass after weathering		
Day 2	Temperature of vinegar		
	Change in mass		
	Percentage change in mass		

LAB **7.1** **INVESTIGATION**

DATA AND OBSERVATIONS, *continued*

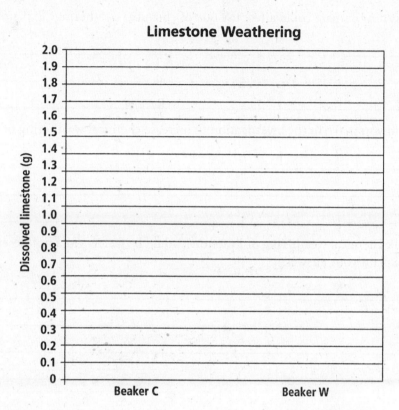

Limestone Weathering

Y-axis: Dissolved limestone (g), from 0 to 2.0

X-axis: Beaker C, Beaker W

ANALYZE

1. What evidence of a chemical reaction did you see when you added vinegar to the limestone chips?

_____

2. What evidence of chemical weathering appeared on the second day?

_____

_____

_____

3. What effect did temperature have on the weathering of the limestone chips?

_____

4. What are some ways to express the amount of change to the limestone due to weathering?

_____

_____

_____

LAB **7.1**   **INVESTIGATION**

CONCLUDE AND APPLY

1. Do the data suggest how climate could affect the rate of chemical weathering? Explain your answer.

_____

_____

_____

2. What changes could you make in the investigation to increase chemical weathering of the limestone?

_____

_____

_____

3. Where do you think chemical weathering of limestone would be faster—in a hot desert or in a rainy northern valley? Explain your answer.

_____

_____

_____

_____

LAB ◆ **7.2** **MAPPING**

Use with
Section 7.3

Global Soils and Climate

Soil is one of our most important natural resources. Humans and other animals depend on plants, which usually grow in soil, for their food and shelter. Except for some mountainous terrain and extremely cold areas on Earth, soil is found everywhere. Soil formation depends on the weathering of bedrock. The type and amount of weathering is closely related to the local climate. Because soils form from different parent bedrock and undergo different climatic conditions, soils vary greatly from one place to another.

PREPARATION

PROBLEM

Are soil types influenced by precipitation and temperature?

MATERIALS

ruler

OBJECTIVES

- **Use** maps to **compare** climate and soils in different regions.
- **Relate** regional soil types to temperature and rainfall.

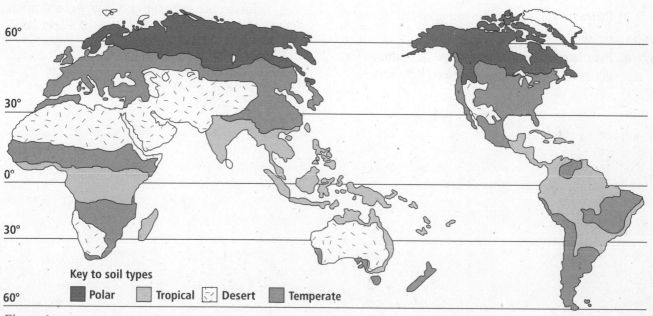

Figure 1

LAB 7.2   **MAPPING**

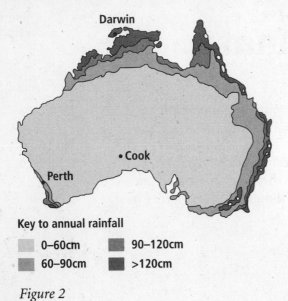

Darwin

• Cook

Perth

Key to annual rainfall

	0–60cm		90–120cm
	60–90cm		>120cm

Figure 2

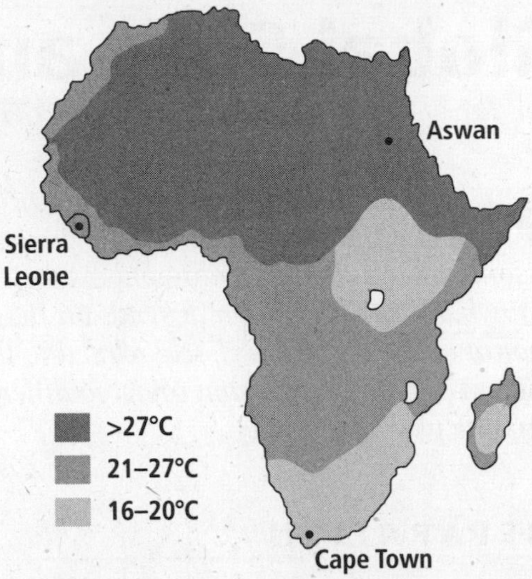

Aswan

Sierra
Leone

	>27°C
	21–27°C
	16–20°C

Cape Town

Figure 3

PROCEDURE

1. Examine Figure 1. The map covers roughly 70°N to 60°S latitude. Notice the key to the four soil types.

2. Estimate the range of latitudes for polar soils, and record your estimate in Table 1.

3. Estimate and record the range of latitudes for tropical soils.

4. Estimate and record the range of latitudes for temperate soils in the northern hemisphere.

5. Examine Figure 2, an annual rainfall map of Australia. In Table 2, record the range of rainfall for the cities of Perth, Cook, and Darwin.

6. Examine Figure 3, a typical temperature day during the summer in Africa. Record the range of temperatures in the cities of Sierra Leone, Aswan, and Cape Town.

LAB 7.2

DATA AND OBSERVATIONS

Table 1

Soil Type	Range of Latitudes
Polar	
Tropical	
Northern temperate	

Table 2

City	Range of Rainfall
Perth	
Cook	
Darwin	
	Range of Temperature
Sierra Leone	
Aswan	
Cape Town	

ANALYZE

1. At what latitudes in the northern and southern hemispheres are desert soils found?

_____

_____

2. What might be the reason for polar soils reaching down to about 40°N in the United States, while in the rest of the world they stay up around 50°N?

_____

_____

3. Compare the rainfall pattern in Australia to the soil types found there.

_____

_____

_____

_____

_____

LAB **7.2** MAPPING

ANALYZE, *continued*

4. Compare the temperature map of Africa with the soil types found in that country.

_____

_____

_____

_____

_____

CONCLUDE AND APPLY

1. Are the soil types found in different areas of Earth related to the climate of that area? Explain your answer.

_____

_____

_____

_____

_____

2. Are the soil types more closely related to temperature or to rainfall?

_____

_____

_____

3. If you had to predict the type of soil found in an area on Earth, what information would you seek?

_____

_____

_____

Use with
Section 8.2

How does wind erosion take place?

Winds transport materials by causing sediment particles to move in different ways. One method of transport is called suspension, in which strong winds cause particles to stay airborne for long distances. Another method of wind transport, called saltation, causes a jumping motion of particles. You will investigate saltation, the most common method of sand transport, when you move sand in a box with the use of a hair dryer.

PREPARATION

PROBLEM

How does wind-blown sand behave around small rocks and other obstructions?

OBJECTIVES

- **Observe** and **compare** the effects of moving air on different particle sizes.
- **Observe** some features of wind deposits.

MATERIALS

hair dryer with cool heat setting
box, cardboard (approximately 30 cm wide, 60 cm long, 10 cm high)
samples of sand and gravel
400-mL beaker
twigs with a number of branches (or steel wool and pencils)
candle
matches
rocks or rubber stoppers of various sizes
protractor
cardboard strip (15 cm × 5 cm)
20-cm thin string
paper clip
tape

SAFETY PRECAUTIONS

- If possible, plug the hair dryer into a ground fault interruptor (GFI) outlet.
- Do not get the hair dryer wet or use it with wet hands.
- Blown sand, clay, or silt can cause injury to the eyes or irritate skin. Do not use the high speed.
- Wear safety goggles and apron during the lab procedure.
- Make sure the heat setting on the hair dryer is set to off if possible.
- Do not stick metal objects into the hair dryer; electrocution can result.
- Wear thermal mitts if the hair dryer becomes hot to the touch.
- Follow your teacher's suggestions for disposing of lab materials.

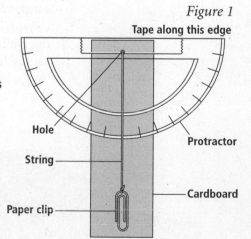

Figure 1

LAB **8.1** **INVESTIGATION**

PROCEDURE

1. First, you will make a type of inclinometer to measure the slope of your sand dunes. With tape, attach the straight side of the protractor to the top of the cardboard strip, as show in Figure 1. Use a straightened paper clip to punch a hole in the cardboard through the hole in the protractor. Thread one end of the string through the hole and tie it. Attach a paper clip to the other end of the string. The paper clip must hang between the bottom of the cardboard and the bottom of the protractor. Set the inclinometer aside.

2. Use the beaker to obtain a sample of sand. Pour enough sand over the bottom of the box to form a layer about 1 cm thick. Distribute the sand evenly over the bottom.

3. Make sure your hands are dry while handling the hair dryer. Plug in and turn on the hair dryer, using the lowest setting. Move the hair dryer so the air is blowing horizontally over the sand. Adjust the speed and/or distance so that sand is moving toward the back of the box. Observe how the sand is moving. Turn off the dryer and record your observations in the table on the next page.

4. Redistribute the sand evenly over the bottom of the box. Place two or three rocks (or rubber stoppers) in the sand. Repeat this step until a sand dune forms. Turn off the dryer and carefully remove the rocks or stoppers from the box.

5. Using the inclinometer constructed in step 1, carefully measure the angle of the sand on the windward and lee-ward sides of the dune (Figure 2). Subtract the reading on the protractor indicated by the hanging string from 90 degrees. Record these values in the table.

Figure 2

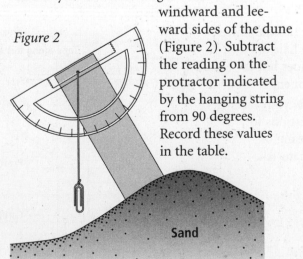

Sand

6. Smooth out the sand in the box. Place a twig in the sand. If you're using steel wool, then pull it apart slightly, so that there is more open space. Anchor the steel wool with a pencil or a small rock. Repeat step 4. Record your observations in the table. Remove the twig and smooth the sand.

7. Invert the beaker and press it into the sand. Place the candle in the sand directly behind the beaker and light it. **CAUTION:** *Make sure the candle is anchored securely in the sand. Do not burn candles near any combustibles.* Hold the dryer at a distance of about 30 cm in front of the beaker. Direct the dryer at the beaker and observe the results. While keeping the dryer at a fixed distance in front of the beaker, back the candle up about 3 cm at a time and observe the results. Turn off the dryer. Record your observations in the table.

8. Remove the beaker and candle and smooth the sand. Invert the beaker and press it into the sand again. Direct the dryer at the beaker. Observe how the sand moves around the beaker. Turn off the dryer. Record your observations in the table.

9. Extinguish the candle and remove it and the beaker from the box. While continuing to operate the dryer, have a partner add more sand very slowly to the box by pouring it from a height of 30 cm. Be sure this is done near the end of the box closest to the hair dryer to avoid blowing the sand out of the box. Record your observations in the table.

10. Remove all but 100 mL of the sand from the box and return it to its container. Put a handful of gravel in the box. Spread the gravel out slightly. Put the sand on top of the gravel, covering it completely. Use more sand if necessary. Direct the hair dryer at the sand. Adjust the speed/distance until the sand is moving slowly. Turn off the dryer and record your observations in the table.

11. Remove the gravel from the box and return it to its container. Put the sand in the beaker and dampen the sand slightly so that it barely sticks together. Pour the wet sand into the center of the box. Direct the dryer at the pile of damp sand and observe what happens. Turn off the dryer and record your observations in the table.

LAB ◆ **8.1** INVESTIGATION

DATA AND OBSERVATIONS

	Observations
Procedure 4	
Procedure 6	Windward side = Leeward side =
Procedure 7	
Procedure 8	
Procedure 9	
Procedure 10	
Procedure 11	
Procedure 12	

ANALYZE

1. Where did the sand dune form relative to any obstacle you placed in the sand?

_____

_____

2. Which side of the dune was steeper? On which side was erosion dominant? On which side was deposition dominant?

_____

_____

_____

3. The twigs and steel wool simulated plants on the surface of the sand. How did they affect the movement of the sand?

_____

_____

_____

4. Describe the burning of the candle as you backed it slowly away from the beaker.

_____

_____

_____

_____

LAB  **8.1**   **INVESTIGATION**

CONCLUDE AND APPLY

1. How would you characterize the force of wind behind an obstruction, and how does this relate to the deposition of sand behind such an obstruction?

_____

_____

_____

_____

_____

2. What could be used to slow the movement of wind-blown sand across a highway?

_____

_____

_____

3. Did the sand movement take place by saltation or suspension? Give evidence for your answer.

_____

_____

_____

LAB ◄ **8.2** ►

DESIGN YOUR OWN

Analysis of Glacial Till

Glaciers deposit two kinds of materials-till and outwash. Till consists of material that was carried by the glacier and deposited as the ice melted. A till deposit contains particles of all sizes in an unlayered mass. Outwash is material deposited by streams flowing from the ice. The moving water sorts the materials so that in an outwash sample, one particle size is more common than other sizes. Outwash deposits usually occur in distinct layers.

PREPARATION

PROBLEM

We want to determine the various particle sizes and what percentage each particle size is of the whole sample. In order to do this, we have to come up with a method of sorting the glacial till and outwash into different particle sizes. We then will measure the relative amount of each particle size in the sample.

OBJECTIVES

- **Determine** the particle sizes in samples of glacial till and outwash.
- **Measure** the relative amounts of each particle size.
- **Compare** the particle sizes and relative amounts of each size in glacial till and outwash.

HYPOTHESIS

Think about the material that might be left behind by a glacier. Would you expect to see more of one particle size than other particle sizes? What would be largest and smallest particle size you might find?

Make a hypothesis about the sizes of particles that will be found in the glacial till and outwash, as well as the percentage of each particle size that will be found in the sample. Write your hypothesis below.

_____

_____

_____

_____

POSSIBLE MATERIALS

unknown glacial till and/or outwash
pan set
balance
beakers
screening sieves
filter paper

SAFETY PRECAUTIONS

- Wear safety goggles and apron during the lab procedure.
- Plan to dispose of wastes as directed by your teacher.

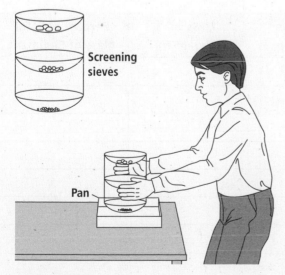

Screening sieves

Pan

LAB **8.2**   **DESIGN YOUR OWN**

PLAN THE EXPERIMENT

- Think about ways in which you can separate the till and outwash. As you look at this till, what differences do you see between the various particles in the till? Can you use these differences to separate the particles? What could you use that would allow the smallest particles to pass through, but keep the larger pebbles behind? What separating tools will you need to separate the various sizes of particles? You will not be able to separate the till or outwash into just one size. Instead, you will have a range of sizes in each group.

- Once you separate the particles into various groups, how will you measure the amount of material in each group? What measuring tools will you need?

- What data will you need to record? Look at the objectives to get a hint. What kind of data table will you construct to make your measurements meaningful? Use or modify the chart below to record your data. Use the bar graph as a model for your data presentation. Use other graphs, as you feel necessary, to make the data more meaningful.

- Use the mesh-size table below as a guide to presenting some of your data.

   Talk with the members of your group until your ideas are well thought out, and then write below the procedures that your team will follow.

_____

_____

_____

_____

_____

_____

_____

_____

_____

_____

_____

_____

_____

_____

_____

_____

_____

_____

DATA AND OBSERVATIONS

Object	Mass of Empty Container (g)	Mass of Container plus Sediment (g)	Mass of Sediment (g)	Percent of Total Mass of Sediment

LAB **8.2** **DESIGN YOUR OWN**

DATA AND OBSERVATIONS, *continued*

MESH SIZES

Top, or largest, screen   =   _____ mm

Second screen from top   =   _____ mm

Third screen from top   =   _____ mm

Bottom, or smallest screen =   _____ mm

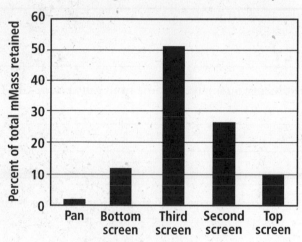

Percentage of Various Sizes in Sample

ANALYZE

1. What method did you use to separate the till and outwash?

_____

_____

2. After separating the till, how many groups did you have?

_____

_____

3. What were the sizes of the particles in each group?

_____

_____

_____

LAB  **8.2**   **DESIGN YOUR OWN**

ANALYZE, *continued*

4. What size particles formed the largest group?

_____

_____

CHECK YOUR HYPOTHESIS

Was your **hypothesis** supported by your data? Why or why not?

_____

_____

CONCLUDE AND APPLY

1. Do you think that the glacial sediment you analyzed came from outwash or till? Explain your answer.

_____

_____

_____

2. Use a chart similar to the one below to categorize your groups' particle sizes as they relate to standardized particle sizes.

0.01 (size in mm)   0.0625   0.125   0.25   0.5   1.0   2.0   4.0   8.0

Fine Silt	Medium Silt	Coarse Silt	Very Fine Sand	Fine Sand	Medium Sand	Coarse Sand	Very Coarse Sand	Granules	Small Pebbles

3. How would you change the investigation if you wanted to measure the outwash in a glacial region of Wisconsin and were going to publish your findings?

_____

_____

_____

_____

LAB **9.1**   **INVESTIGATION**

Use with
Section 9.1

Analyzing Watersheds

A *watershed is the land area whose water drains into a particular stream system. Any pollutants from the surface or in the groundwater will find their way into wells, surface streams, and lakes in a watershed. A watershed is also affected by any construction that disrupts its drainage pattern. You can gauge the health of a watershed by looking at condition indicators and vulnerability indicators. A condition indicator represents present conditions, such as contaminated sediments and groundwater that contains chemicals. Vulnerability indicators represent conditions that may adversely affect the watershed in the future, such as large human populations and the potential for agricultural runoff.*

PREPARATION

PROBLEM
Determine the health of a watershed by analyzing indicators and establish goals to improve the health of the watershed.

OBJECTIVES
- **Examine** maps showing condition and vulnerability indicators.
- **Analyze** maps and **establish** a watershed health report.

- **Develop** a list of goals aimed at reversing damage and improving the health of the watershed.

MATERIALS
ruler
colored markers

PROCEDURE

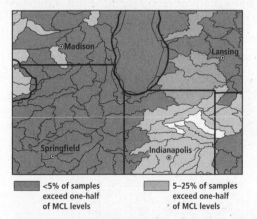

Figure 1. Chemicals from many different sources in the groundwater. Before the water is used, many of these chemicals will be filtered out or neutralized. (1997)

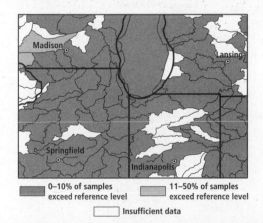

Figure 2. Copper, chromium, nickel, and zinc in the groundwater. Consumption of high concentrations of these chemicals causes illness or death. (1997)

LAB ◇ **9.1**   **INVESTIGATION**

PROCEDURE, *continued*

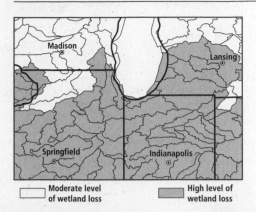

Figure 3. Wetlands lost because of human use. Wetlands make important contributions to the health of a watershed by purifying water, filtering runoff, abating floods, and decreasing erosion. (1997)

Legend: Moderate level of wetland loss | High level of wetland loss

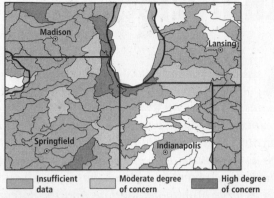

Figure 4. Chemicals found in bottom sediments. These chemicals can harm or kill bottom-dwelling organisms. They can also accumulate in organisms and move up the food chain to fish, shellfish, and humans. (1997)

Legend: Insufficient data | Moderate degree of concern | High degree of concern

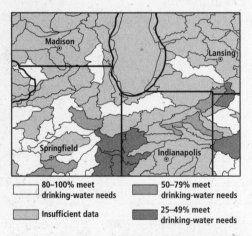

Legend: 80–100% meet drinking-water needs | 50–79% meet drinking-water needs | Insufficient data | 25–49% meet drinking-water needs

Figure 5. Watersheds that meet drinking-water needs of a human population. (1997)

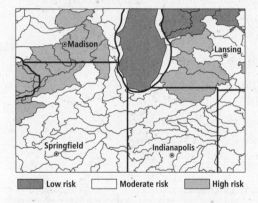

Legend: Low risk | Moderate risk | High risk

Figure 6. Fish advisories issued for the watershed. Advisories indicate the accumulation of toxic substances in fish and shellfish, making them unsafe for human consumption. (1997)

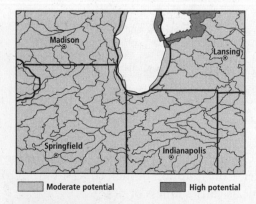

Legend: Moderate potential | High potential

Figure 7. Pesticide runoff from farms. Watersheds with high scores are at greater risk of contamination of surface water by pesticides. (1997)

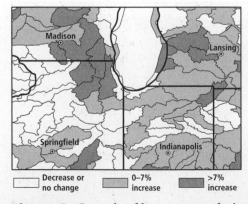

Legend: Decrease or no change | 0–7% increase | >7% increase

Figure 8. Growth of human population. Population increases can result in increased pollution of the water. (1997)

LAB 9.1   **INVESTIGATION**

PROCEDURE, *continued*

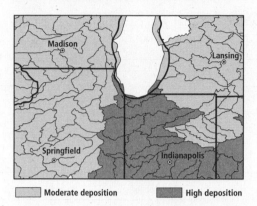

Figure 9. Atmospheric deposition of nitrogen. Nitrogen and phosphorus are nutrients that can cause algal and cyanobacterial blooms and other problems in surface water and groundwater. (1997)

1. Examine the six condition indicators (Figures 1–6) and four vulnerability indicators (Figures 7–10). Each irregular area on the map represents a watershed.

2. Pick one watershed and draw around its perimeter on each map. This will be the watershed that you examine for each indicator.

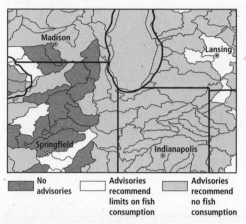

Figure 10. Nitrate risk from various sources, such as fertilizers, atmospheric deposition, and karst aquifers. (1997)

3. Analyze each condition indicator. In Figure 11, draw a bar for each condition indicator to represent your interpretation of the map.

4. Analyze each vulnerability indicator. Draw a bar for each vulnerability indicator in Figure 12.

DATA AND OBSERVATIONS

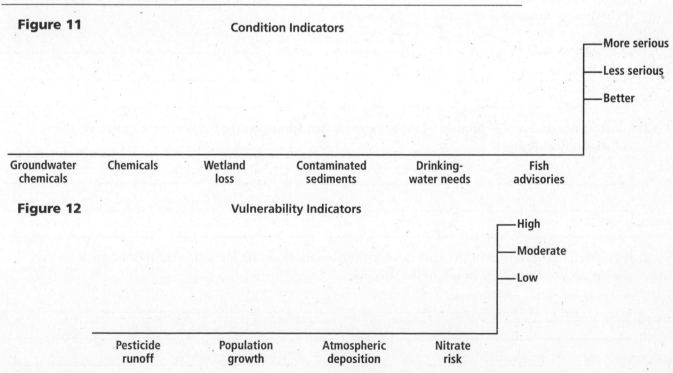

Figure 11   **Condition Indicators**

Figure 12   **Vulnerability Indicators**

LAB 〈 **9.1** 〉 **INVESTIGATION**

ANALYZE

1. Write a short narrative describing the present condition of the watershed that you chose.

_____

_____

_____

2. Describe the vulnerability of the watershed.

_____

_____

_____

3. Are the watershed conditions similar to those of surrounding watersheds? Why do you suppose they are or are not?

_____

_____

_____

_____

CONCLUDE AND APPLY

1. Taking into consideration the many pressures on the watershed you chose, predict what will happen if conditions remain the same for the next 50 years.

_____

_____

_____

2. What is your assessment of most of the watersheds that appear on the maps of condition and vulnerability indicators?

_____

_____

_____

3. If you were the governor of the state in which these watersheds are located, what specific goals would you set up to improve the health of the citizens?

_____

_____

_____

LAB 9.2 MAPPING

Use with
Section 9.2

Interpreting a River's Habits

All *stream systems generally start from rain running off the land. A stream develops further, depending on the amount of available water, the slope of the land, and the underlying type of bedrock. Fast-moving streams follow a straighter path than do slow-moving streams, which tend to form meanders. Oxbow lakes often form from meandering streams and rivers. Below is a topographic map of the Souris River valley in north-central North Dakota. This area was under a continental glacier during the ice ages. The surface is largely covered with moraine deposits.*

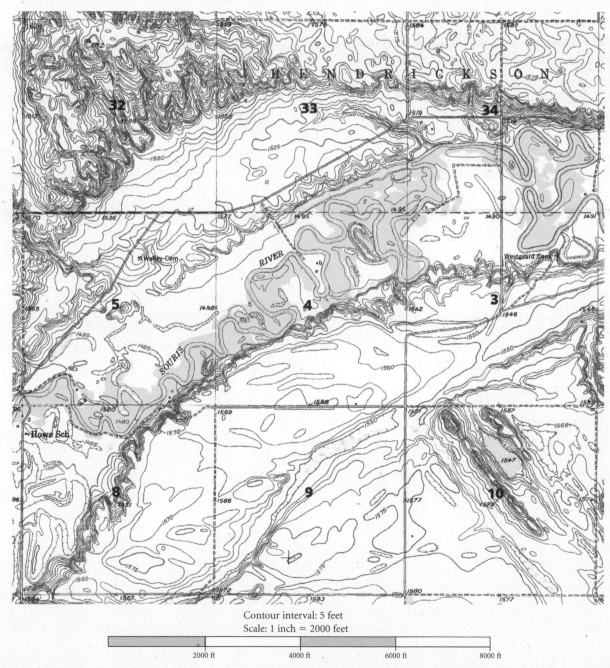

Contour interval: 5 feet
Scale: 1 inch = 2000 feet

2000 ft   4000 ft   6000 ft   8000 ft

LAB **9.2** **MAPPING**

PREPARATION

PROBLEM
What can a topographic map tell about
a river and its surroundings?

OBJECTIVES
Use a topographic map to **answer**
questions about a river and its valley.

MATERIALS
ruler

PROCEDURE

1. The topographic map has a contour interval of
5 feet. The scale is 1 inch for 2000 feet. Study
the map and answer questions 1–4 in the table.

2. The river drops about 2 feet in elevation across
the map. Determine the gradient and answer
questions 5–7.

3. Examine the floodplain of the river. Notice
that the contour lines along the river run into
one another. This indicates that natural levees
occur and that at some places they are at least
5 feet high. Answer questions 8–10.

4. Notice that there are numerous elongated
depressions in the floodplain. Answer
question 11.

5. Examine the structures across the top of
the map in sections 32, 33, and 34. Answer
questions 12 and 13.

LAB ◁ **9.2** ▷   **MAPPING**

DATA AND OBSERVATIONS

Table

Question	Answer
1. What is the approximate difference between the lowest point and the highest elevation?	
2. What is the straight-line distance from where the river enters the map to where it leaves the map?	
3. What is the approximate length of the river's course that you can see on the map?	
4. What does the difference between the two distances in questions 2 and 3 tell you about the river's gradient?	
5. What is the river's gradient per 100 feet?	
6. What is the river's gradient per mile?	
7. Estimate the rate at which the river flows: very slowly, slowly, or rapidly.	
8. Approximately how wide is the floodplain of the river?	
9. What happened just east of Westgaard Cemetery in section 3?	
10. Has what you described in question 9 occurred anywhere else on the map? If so, where?	
11. What are the depressions in the floodplain called?	
12. Is there any evidence of erosion in sections 32, 33, and 34?	
13. Is there any evidence of other stream valleys?	

LAB **9.2** **MAPPING**

ANALYZE

1. How is a structure like the one identified in question 9 formed?

_____

_____

_____

2. What do you suspect is the origin of the generally flat land between the present floodplain bluff and the steep bank in the northern third of section 33 and most of section 32?

_____

_____

_____

CONCLUDE AND APPLY

1. From your interpretation of the topographic map, describe the Souris River's shape, flow rate, and amount of downcutting into the bedrock.

_____

_____

_____

_____

2. Describe the geography around the Souris River.

_____

_____

_____

3. What do you think the overall rain pattern in this area might be: little rain, moderate, or heavy? Support your answer with an explanation.

_____

_____

_____

_____

LAB 10.1   **INVESTIGATION**

Measuring Permeability Rate

After a rain, you may have noticed that puddles are left on the sidewalks, but not on the grass next to them. If the same amount of rain falls on both surfaces, why does more water remain on one surface? A lawn is usually much more permeable than a cement or asphalt sidewalk. Permeability is the ability of a material to let water pass through it.

PREPARATION

PROBLEM
How does the water permeability of different soil components vary?

OBJECTIVES
- **Measure** the water permeability of various types of soil.
- **Compare** and **contrast** the permeability of pure and mixed materials.

MATERIALS
hand lens
100 mL sand
100 mL pebbles
100 mL potter's clay
100 mL unsorted soil
100-mL graduated cylinder

water
stopwatch
4 rubber bands
4 cheesecloth squares
4 large funnels
500–1000-mL beakers (4)

SAFETY PRECAUTIONS

- Potter's clay that is airborne can irritate eyes and nose; wear goggles while handling dry clay.
- Wipe up any spills immediately.
- Wear an apron during the lab procedure to avoid staining your clothing.

LAB 10.1   **INVESTIGATION**

PROCEDURE

1. Examine the sand, pebbles, clay, and unsorted soil with a hand lens. Look for differences in the particle size as well as other observable characteristics. Record this information under Observations.

2. Line four funnels with squares of cheesecloth. Secure the cheesecloth with a rubber band. Set each funnel on top of a beaker.

3. Put 100 mL of sand in one funnel, 100 mL of pebbles in a second, 100 mL of clay in a third, and 100 mL of unsorted soil in a fourth. Leave at least 3–4 cm of space above the material in each funnel.

4. Pour water through the funnel until the water just starts to drip into the beaker. Stop pouring the water and wait until the water stops dripping. Empty the water out of the beaker.

5. Slowly pour 100 mL of water into the funnel containing the sand. Do not let the water overflow the funnel. Start the stopwatch when the water begins to drip out of the funnel.

6. Stop the stopwatch when the water stops dripping out of the funnel or after 5 minutes. Record the time to the nearest second in the table below.

7. Measure the water that drained into the beaker and record the amount.

8. Repeat steps 4–7 for the pebbles and clay.

9. Calculate the permeability of the three materials by dividing the amount of water that drained from each funnel by the drainage time in seconds. Express the value as milliliters per second.

10. Based on the permeability of the first three materials, estimate and record the permeability of the unsorted material.

11. Repeat steps 4–7 for the unsorted material. Calculate the permeability.

12. Do not put the wet materials in the sink or trash. Store them where your teacher directs you.

DATA AND OBSERVATIONS

OBSERVATIONS

_____

_____

_____

_____

_____

_____

_____

	Sand	Pebbles	Clay	Unsorted
Time for draining (s)				
Drained water (mL)				
Estimated permeability rate (mL/s)				
Calculated permeability rate (mL/s)				

LAB **10.1** **INVESTIGATION**

ANALYZE

1. Create a bar graph for your data, using the empty graph below.

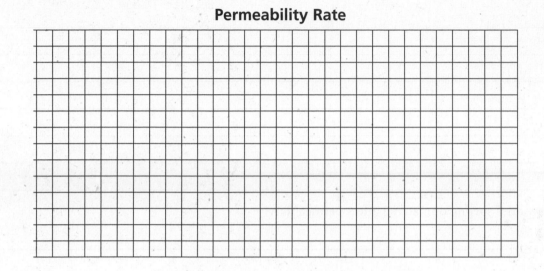

Permeability Rate

Permeability (mL/sec)

Material

2. Compare the permeability rate of the four materials.

_____

_____

_____

3. Did you accurately predict the permeability rate through the unsorted soil? If not, why not?

_____

_____

LAB ◄ **10.1** ►   **INVESTIGATION**

CONCLUDE AND APPLY

1. Based on your observation of each sample, suggest an explanation for the differences in their permeability.

_____

_____

_____

_____

_____

_____

_____

2. How does sorting of material affect permeability?

_____

_____

3. Based on the results of this investigation, would you expect to get more water from a well dug in sand, pebbles, clay, or unsorted material? Give reasons for your answer.

_____

_____

_____

_____

LAB ◆ **10.2** **DESIGN YOUR OWN**

Use with
Section 10.3

Analysis of Drinking Water

In many areas of the United States, groundwater is the only economical source of household water. If groundwater becomes contaminated, it can be cleaned up only with difficulty and at great expense. The contaminants originate above ground and often result from human activities. Wells bring groundwater up and are a source of drinking water and a direct pathway from the land surface into the water supply. Well water should be tested at least once a year. If the well is in an area of potential pollution sources such as farms, landfills, and toxic disposal sites, it should be tested periodically.

PREPARATION

PROBLEM
How can you tell if drinking water from a well is safe?

OBJECTIVES
- **Examine** test results for well water.
- **Interpret** the test results and **assess** health risks.
- **Write** a report describing test results, listing health risks, and making recommendations.

HYPOTHESIS
As a group, write a hypothesis about the safety of the well water.

_____

_____

_____

POSSIBLE MATERIALS
paper
computer

PLAN THE EXPERIMENT

Study Table 1, listing contaminants, their maximum contaminant level (MCL), health risks to humans, possible sources of the pollutants, and possible treatments to clean the water. The MCL is the maximum contamination allowed by the Environmental Protection Agency (EPA). The suggested water treatments would be fairly cheap to do

on site and do not affect the water in the well or in the ground. Examine the well test results (Table 2), and determine any risk to humans. Design and write a report of your discoveries. Include health risks and other information and suggestions that you think are necessary. Keep your report about a page long.

LAB 10.2   **DESIGN YOUR OWN**

DATA AND OBSERVATIONS

Table 1

Contaminant	MCL (mg/L)	Risks to Humans	Contaminant Sources	On-Site Treatment
Arsenic	0.05	Weight loss, depression, cancer	Pesticides, improper waste disposal, mining	Reverse osmosis, filtration, distillation
Barium	2	Toxic to heart, blood vessels, and nerves	Paints, diesel-fuel combustion, mining	Reverse osmosis, filtration, distillation
Benzopyrene	0.0002	Cancer	Coal-tar coating, fossil fuels	Activated carbon
Cadmium	0.005	Kidney damage, mutations	Fertilizers, sewage, discarded batteries	Reverse osmosis, filtration, distillation
Chromium	0.1	Lung tumors, nervous-system damage, accumulation in organs	Septic systems, industrial discharge, mining sites	Reverse osmosis, filtration, distillation
2-4 D	0.07	Cancer, liver and kidney damage	Herbicides, aquatic-weed control	Activated carbon
Dioxin	0.00000003	Cancer, mutations	Impurity in herbicides, chemical by-products	Activated carbon
Cyanide	0.2	Thyroid and nervous-system damage	Fertilizer; mining; electronics, steel, and plastic manufacturing	Ion exchange, reverse osmosis, chlorination
Lead	0.015	Reduced mental capacity, neurological problems	Paint, diesel-fuel combustion, discarded batteries, old paints and solder	Activated carbon, ion exchange, reverse osmosis
Methoxychlor	0.04	Reduced growth; impact on liver, kidneys, and nerves	Insecticide for fruits, vegetables, alfalfa, livestock, pets	Activated carbon
Nitrate	10	Blue-baby disease	Livestock facilities, septic systems, fertilizers	Ion exchange, distillation, reverse osmosis
Picloram	0.5	Kidney and liver damage	Herbicide on broadleaf plants	Activated carbon
Thallium	0.002	Skin irritation	Electronics, glass, and pharmaceuticals manufacturing	Ion exchange, distillation

LAB 10.2   **DESIGN YOUR OWN**

DATA AND OBSERVATIONS, *continued*

Table 2

Results for Well 1		Results for Well 2	
Contaminant	Amount (mg/L)	Contaminant	Amount (mg/L)
Arsenic	0.045	Arsenic	0.06
Cadmium	0.007	Barium	3.0
2-4 D	0.01	Benzopyrene	0.0000
Lead	0.017	Cadmium	0.001
Methoxychlor	0.04	2-4 D	0.00
Nitrate	0.01	Dioxin	0.00000000
Picloram	0.000	Lead	0.000
Thallium	0.000	Nitrate	0.00

ANALYZE

1. What information did you provide to the well owner and why?

_____

_____

_____

_____

2. What criterion did you use to determine if a health warning was necessary?

_____

_____

_____

_____

3. Did you suggest additional water tests? Why or why not?

_____

_____

_____

_____

_____

LAB **10.2** **DESIGN YOUR OWN**

ANALYZE, *continued*

4. Which water treatments did you suggest, and why?

_____

_____

_____

_____

_____

_____

CHECK YOUR HYPOTHESIS

Was your **hypothesis** supported by your data? Why or why not?

_____

_____

CONCLUDE AND APPLY

1. What did you conclude about the water from wells 1 and 2?

_____

_____

_____

2. If you had more room on the report, what additional information would you provide?

_____

_____

_____

3. If you were the EPA director, what directives would you make to warn people of possible health risks from their well water?

_____

_____

_____

LAB ◄ **11.1** ►   **INVESTIGATION**

Temperature Inversion

In some cities, the weather report often warns of high air-pollution levels. People are asked not to drive unless it is absolutely necessary, and open fires and barbecues are forbidden. A frequent reason for high levels of air pollution near the ground is a temperature inversion in the atmosphere. Although temperature and pressure in the overall troposphere decrease with height, the temperature inversion is an exception to this rule.

PREPARATION

PROBLEM

How can you detect a temperature inversion, and how does it trap pollution?

MATERIALS

ruler
calculator

OBJECTIVES

- **Graph** temperature data for the atmosphere.
- **Describe** how a temperature inversion affects ground-level pollution.

PROCEDURE

1. Use Box 1 to graph data sets A and B. The horizontal axis will be height and the vertical axis will be temperature. Label these axes.

2. Look at the data sets and choose suitable ranges and intervals for the axes. Mark the axes accordingly.

3. Plot data set A on your graph. Connect the points with a solid line. Plot data set B on the same graph but connect the points with a dashed line.

4. Indicate on your graph which line represents which data set. Give the graph a title.

5. When air is heated, it expands. As air expands, the number of molecules in a particular volume,

for example, 1 m³, decreases. So the mass of air molecules per cubic meter—or its density—decreases. Because of this relationship, if the pressure of the air remains unchanged, then its density is inversely proportional to its temperature, provided the temperature is expressed in kelvins (K). The Kelvin scale starts at absolute zero, which corresponds to −273.15°C. To convert a temperature in degrees Celsius to kelvins, you simply add 273.15. Convert the temperatures in the table into kelvins.

6. You can work out the density of air at a particular height from the temperature data if you know the density of air at ground level.

PROCEDURE, *continued*

Density is inversely proportional to absolute temperature; therefore, the density at height *x* is equal to the density at ground level times the absolute temperature (kelvins) at ground level divided by the absolute temperature at height *x*. Use this information to calculate the density at each height for both data sets. Assume that the air density at ground level in both data sets is 1 kg/m³.

7. Use Box 2 to make a graph of density versus height similar to the first graph you made.

DATA AND OBSERVATIONS

Table

| Height (m) | Data Set A | | | Data Set B | | |
	Temperature (°C)	Absolute Temperature (K)	Density (kg/m³)	Temperature (°C)	Absolute Temperature (K)	Density (kg/m³)
0	20.0			20.0		
100	19.5			19.3		
200	18.7			18.8		
300	18.0			18.0		
400	17.5			19.0		
500	16.9			19.5		
600	16.0			19.4		
700	15.5			18.8		

Box 1

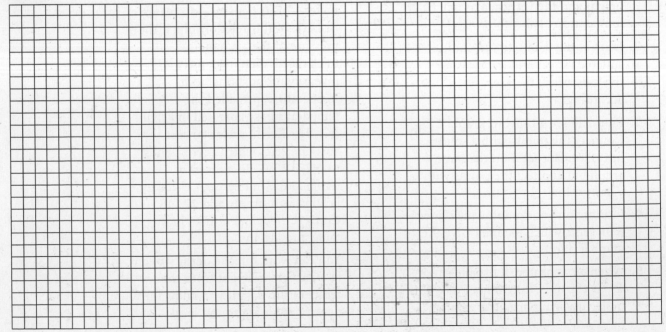

LAB **11.1** | **INVESTIGATION**

Box 2

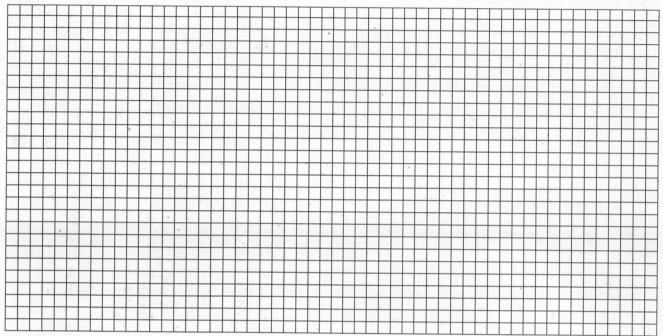

ANALYZE

1. For data set A, does the temperature increase or decrease as height increases? At what altitude does the temperature first change by 1°C?

   _____

2. At what point do the two plotted lines from data sets A and B intersect?

   _____

   _____

3. Describe the plotted data of set B above 300 m.

   _____

   _____

4. Which data set shows normal conditions and which shows a temperature inversion?

   _____

   _____

5. In the data set with the temperature inversion, use your graph in Box 2 to compare the density in two regions: 300–500 m and 500–700 m.

   _____

   _____

LAB ◁ **11.1** ▷ **INVESTIGATION**

CONCLUDE AND APPLY

1. Air pollutants tend to move from more dense regions toward less dense regions. What does this imply for the movement of air pollutants in the data set with the temperature inversion?

_____

_____

_____

_____

2. At what height would you expect to find the greatest concentration of air pollutants in the data set with the temperature inversion?

_____

_____

3. In your own words, summarize how temperature inversions increase air pollution at ground level.

_____

_____

_____

_____

_____

_____

What is in the air?

Air is mostly composed of nitrogen and oxygen, with varying amounts of water vapor. However, air also contains particulates like dust, smoke, and pollen. It can be very useful to know just what particulates are in the air, especially for health reasons. One source of data is the daily pollen count; pollen affects hay fever and asthma sufferers. The concentration of particulates varies with time and weather. Health and safety authorities also monitor particulates in workplaces and schools.

PREPARATION

PROBLEM

What is the particulate content of the air? How does it change from day to day?

OBJECTIVES

- **Observe** how the daily weather affects the number and type of particulates in the air.
- **Research** the number and type of particulates in the air around your school.

HYPOTHESIS

As a group, discuss the atmospheric changes caused by different weather patterns. Write a hypothesis about how the weather affects the concentration of particulates outside and inside your school.

_____

_____

_____

_____

POSSIBLE MATERIALS

coffee filters
rubber band
thermometer
microscope
vacuum cleaner with intake hose
outdoor extension cord
masking tape
5 microscope slides
petroleum jelly
5 petri dishes with lids

SAFETY PRECAUTIONS

- Wear safety goggles and an apron during the lab procedure.
- Do not use the vacuum cleaner outside in inclement weather. Keep the cord away from water.

 **LAB 11.2**   **DESIGN YOUR OWN**

PLAN THE EXPERIMENT

Review the list of possible materials. Plan how to monitor the daily variation of the atmosphere's particulate content both outside and inside the school during a particular period of time. Choose the sites you will monitor. How will you determine the concentration of particulates in the air each day at each location? What daily location measurements will you make? Plan to collect a week's worth of data. Design a table to record your data. Outline your plan and have your teacher approve it before you begin the experiment.

_____

_____

_____

_____

_____

_____

_____

_____

_____

_____

_____

_____

_____

_____

_____

_____

LAB 11.2

DESIGN YOUR OWN

DATA AND OBSERVATIONS

DATA TABLE

LAB **11.2**   **DESIGN YOUR OWN**

ANALYZE

1. Which locations had the highest particle count? Which had the lowest?

_____

_____

2. What types of particles did you collect outdoors?

_____

3. Which elements of the weather most clearly affected the number of particles collected outdoors?

_____

4. Did any of your indoor locations show the same trend of variation as the outdoor locations? If so, which ones?

_____

_____

5. Did any of your indoor measurements vary with the weather in a significantly different way from the outdoor counts? Explain your answer.

_____

_____

_____

CHECK YOUR HYPOTHESIS

Was your **hypothesis** supported by your data? Why or why not?

_____

_____

CONCLUDE AND APPLY

1. Based on your research, is a person with allergies more likely to have trouble indoors or outdoors at your school? Explain your answer.

_____

_____

2. Under what circumstances would you expect a pollen warning to be issued?

_____

_____

LAB **12.1**   **INVESTIGATION**

Use with
Section 12.2

Modeling the Coriolis Effect

The rotation of Earth in an easterly direction causes the Coriolis effect. The Coriolis effect, in turn, influences the direction of all free-moving objects, such as air and water. For example, in the northern hemisphere, air moving from the north pole toward the equator is deflected to the right. In the southern hemisphere, air moving from the south pole toward the equator is deflected to the left. The Coriolis effect greatly influences the movement of global wind patterns and ocean currents.

PREPARATION

PROBLEM

How does the Coriolis effect deflect the movement of air and water in each hemisphere?

OBJECTIVES

- **Model** the Coriolis effect in the northern and southern hemispheres.
- **Sketch** various movements caused by the Coriolis effect.

- **Infer** how the Coriolis effect influences global wind patterns and ocean currents.

MATERIALS

globe
red, blue, yellow, and green chalk

PROCEDURE

1. Working with a partner, place a globe on a steady, flat surface. Locate the equator, the north pole, and the south pole on the globe.

2. Have your partner rotate the globe in a counterclockwise direction at a slow, steady speed. As the globe rotates, use blue chalk to draw a line from the north pole to the equator. Sketch the line in circle A on the next page. Mark the four compass directions and the equator.

3. Use red chalk to draw a line from the equator to the north pole while your partner continues to rotate the globe. Sketch this line in circle B. Then add compass directions and the equator.

4. As your partner rotates the globe in a counterclockwise direction, use green chalk to draw a line from the south pole to the equator. Sketch the line in circle C. Then add compass directions and the equator.

5. As your partner rotates the globe in a counterclockwise direction, use yellow chalk to draw a line from the equator to the south pole. Sketch the line in circle D. Then add compass directions and the equator.

LAB 12.1 INVESTIGATION

DATA AND OBSERVATIONS

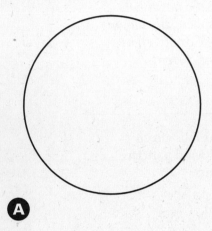

A

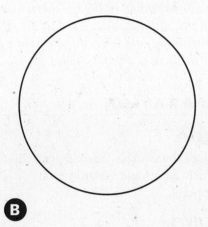

B

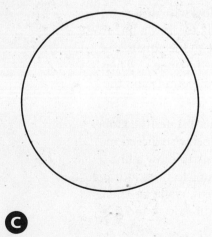

C

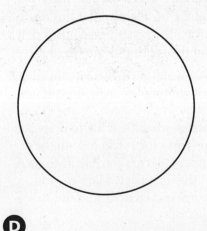

D

LAB **12.1**   **INVESTIGATION**

ANALYZE

1. Compare your four sketches. Describe any patterns that you observe.

_____

_____

_____

2. Why was it necessary to rotate the globe in a counterclockwise direction?

_____

_____

3. Suppose that each line represents a wind system. Describe where each originates and in which direction it moves.

_____

_____

_____

_____

_____

_____

4. Ocean surface currents, which affect weather and climate, move in circular patterns. Use sketches A and C to infer the direction in which these currents move in the northern and southern hemispheres.

_____

_____

_____

LAB **12.1** **INVESTIGATION**

CONCLUDE AND APPLY

1. Summarize how freestanding objects are deflected by the Coriolis effect in the northern and southern hemispheres.

_____

_____

_____

_____

2. How would these objects move if the Coriolis effect did not exist? Explain your answer.

_____

_____

_____

_____

LAB ◁ **12.2** ▷ **DESIGN YOUR OWN**

Use with Section 12.4

Predicting the Weather

Accurate weather forecasts depend on a variety of factors that can change by the minute, such as temperature, wind speed and direction, relative humidity, and air pressure. In addition, meteorologists must incorporate data from the lower atmosphere and the upper atmosphere when they make forecasts. Because so many factors affect weather—and because these factors constantly change—predicting the weather is a complex task.

PREPARATION

PROBLEM

How can you make an accurate weather forecast?

OBJECTIVES

- **Analyze** and **interpret** weather data from a variety of sources.
- **Measure** and **record** weather conditions.
- **Predict** the weather.
- **Determine** the accuracy of weather forecasts.

HYPOTHESIS

As a group, discuss what sources you will use to predict the weather. You may want to use the Glencoe Science Web Site, newspapers, television, radio, or your own weather observations. Write a general prediction of the weather for the next 5 days.

_____

_____

_____

_____

POSSIBLE MATERIALS

weather maps
alcohol-based thermometer
barometer
anemometer
wind vane
rain gauge

SAFETY PRECAUTIONS

- Be careful when handling weather instruments. If a thermometer or other glass instrument should break, do not attempt to clean it up. Notify your teacher immediately.
- Avoid using mercury-based thermometers. Mercury is toxic.
- Wear safety goggles during the lab procedure.

LAB  12.2   DESIGN YOUR OWN

PLAN THE EXPERIMENT

As a group, decide how you will make your weather forecast. Will you use television, Internet, radio, or newspaper weather reports? Will you gather your own weather data? Or will you use a combination of sources? Assign tasks to each member of your group. For example, who will be responsible for measuring or obtaining data about each of the weather conditions in Tables 1 and 2? Who will summarize your data and judge the accuracy of the predictions? Have your teacher approve your plan before you begin.

Use Table 1 to record your weather predictions for the next 5 days. Make a prediction for each of the weather elements listed in the table. Then use Table 2 to record actual weather conditions for each day.

DATA AND OBSERVATIONS

Table 1

Predicted Weather					
	Day 1	**Day 2**	**Day 3**	**Day 4**	**Day 5**
Temperature					
Wind speed					
Wind direction					
Cloud cover					
Precipitation					
Presence of high or low pressure					

LAB 12.2   DESIGN YOUR OWN

DATA AND OBSERVATIONS, *continued*

Table 2

Actual Weather					
	Day 1	**Day 2**	**Day 3**	**Day 4**	**Day 5**
Temperature					
Wind speed					
Wind direction					
Cloud cover					
Precipitation					
Presence of high or low pressure					

ANALYZE

1. What sources did you use to make your weather predictions? Be specific.

_____

_____

_____

2. Which weather conditions were most helpful in terms of making a forecast? Which weather conditions were least helpful?

_____

_____

_____

 **LAB 12.2**   **DESIGN YOUR OWN**

ANALYZE, *continued*

3. Were your predictions for some weather conditions more accurate than others? Explain your answer.

_____

_____

_____

4. Compare your predictions to those of other groups. Was one source consistently more accurate than others? If so, why was it?

_____

_____

_____

CHECK YOUR HYPOTHESIS

Was your **hypothesis** supported by your data? Why or why not?

_____

_____

CONCLUDE AND APPLY

1. Summarize your data. Describe how the accuracy of your forecast changed over time.

_____

_____

_____

2. As a group, brainstorm ways to improve the accuracy of your forecast. Record your ideas.

_____

_____

_____

3. Based on your results, infer why long-term weather forecasts are less reliable than short-term weather forecasts.

_____

_____

_____

 **LAB** ◆ **13.1** **INVESTIGATION**

**Use with
Section 13.4**

Observing Flood Damage

Floods are the main cause of thunderstorm-related deaths in the United States each year. Floods can happen in a number of ways. Sometimes rain falls over a large area and drains into a river faster than the water can move downstream. The river overflows its banks and floods low-lying areas surrounding it. Floods can also occur when large amounts of rain fall on asphalt, concrete, or other surfaces that cannot easily absorb water. Yet another type of flooding, a flash flood, occurs when torrential rains cause a sudden overflow of a river or other water-drainage feature.

PREPARATION

PROBLEM
How do floods affect rates of erosion and human-built structures?

OBJECTIVES
- **Model** different types of floods.
- **Observe** and **record** rates of erosion.
- **Determine** how floods affect local communities.
- **Discuss** ways to reduce flood damage.

MATERIALS
stream table
sand
water source
hose
metric ruler
wooden sticks
large sheet of plastic

SAFETY PRECAUTIONS

- Wear safety goggles and an apron during the lab procedure.
- Be careful when using the hose. To avoid getting shocked, do not use the hose near an electrical receptacle.
- If spills occur, wipe them up immediately to prevent accidents.

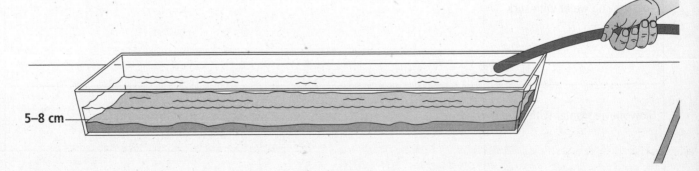

5–8 cm

 **LAB 13.1**

 **INVESTIGATION**

PROCEDURE

1. As a group, set up the stream table on a long, low table. Carefully pour sand into the stream table. The sand should cover the bottom to a depth of 5 to 8 cm.

2. Attach the hose to the water source. Start the water flowing slowly and observe any erosion that occurs. Stop the water and record your observations in the table.

3. Smooth out the sand so that it once again covers the bottom of the stream table to a depth

of at least 5 cm. Insert a wooden stick, which represents a bridge support, in the middle of the stream table. Repeat step 2.

4. Remove the wooden stick and smooth out the sand. Repeat steps 2 and 3 with the water flowing rapidly.

5. Remove the wooden stick and smooth out the sand. Cover the sand with the sheet of plastic and repeat step 2.

DATA AND OBSERVATIONS

Conditions	Observations
Slow-moving water	
Slow-moving water with stick	
Fast-moving water	
Fast-moving water with stick	
Slow-moving water with plastic	

LAB 13.1   **INVESTIGATION**

ANALYZE

1. Compare and contrast the flow of the water in steps 2 through 5.

_____

_____

_____

_____

_____

2. When the wooden stick was in the sand, in what parts of the stream did the water flow fastest and slowest?

_____

_____

_____

3. During which step did most erosion occur? How could you tell?

_____

_____

_____

_____

4. How could you revise this lab to increase the duration of the model flood?

_____

_____

_____

5. Which step in the procedure modeled a flash flood? Which step modeled a flood in an urban area? Explain your answers.

_____

_____

_____

_____

_____

LAB ◁ **13.1** ▷   **INVESTIGATION**

CONCLUDE AND APPLY

1. Based on your observations, why do bridges become unsafe during floods?

   _____

   _____

   _____

   _____

   _____

2. What kind of damage might a flash flood cause to houses along a riverbank?

   _____

   _____

   _____

   _____

   _____

3. As a group, brainstorm ways of reducing flood damage to homes and other human-built structures. Record your ideas here.

   _____

   _____

   _____

   _____

   _____

   _____

LAB 13.2 **DESIGN YOUR OWN**

Use with
Section 13.3

Building Hurricane-Proof Homes

Hurricanes are the most powerful of all storms, so they can cause tremendous damage to property and lives. Much of this damage is a result of violent winds that can exceed 250 km/h. Winds of more than 60 km/h can affect areas as far as 400 km from the center of the storm. In the United States, the National Weather Service issues hurricane warnings that help reduce loss of lives. Before a hurricane strikes a particular area, people temporarily move out of the path of the storm. Buildings, however, cannot be moved. Many are damaged or destroyed by high winds. Scientists and engineers are continually working to design stronger buildings that can withstand hurricane-strength winds.

PREPARATION

PROBLEM

What sort of structure can best withstand hurricane-strength winds?

OBJECTIVES

- **Design** and **construct** a model home using an assortment of building materials.
- **Test** the strength of the structure in a model hurricane.
- **Analyze** different structural designs and **infer** which would best withstand severe storm conditions.

HYPOTHESIS

As a group, discuss how you could use the materials provided by your teacher to design and build a model home. Form a hypothesis about the factors that might affect the ability of your structure to withstand a model hurricane. Write your hypothesis below.

_____

_____

_____

_____

POSSIBLE MATERIALS

wooden sticks
toothpicks
cardboard squares of various sizes
modeling clay
marshmallows
scissors
tape
glue
fan with variable speeds

SAFETY PRECAUTIONS

- Be careful when using the scissors and the electric fan.
- Follow your teacher's suggestions for disposing of lab materials.
- Wear safety goggles and an apron during the lab procedure.

 **LAB** **13.2**   **DESIGN YOUR OWN**

PLAN THE EXPERIMENT

Review the list of possible materials. Working in pairs, design a building that could withstand hurricane-strength winds. Sketch and label the design in the blank space provided below. Share your design with your group and decide on the best design. You might incorporate elements from several designs. Then, sketch and label the revised plan. Discuss how you will test your design. What will your control be? How will you model the hurricane? How will you vary hurricane conditions? How many times should you repeat the test? Construct your model. When your model is complete, test its strength. Draw a data table to record your observations. Have your teacher approve your plan before you begin.

_____

_____

_____

_____

_____

_____

_____

_____

_____

DATA AND OBSERVATIONS

DESIGNS AND DATA TABLE

LAB 13.2   **DESIGN YOUR OWN**

ANALYZE

1. Describe your model. Explain the function of each part.

_____

_____

_____

2. Explain how you modeled hurricane-strength winds. How did you determine the strength of your model building?

_____

_____

_____

_____

3. Did your building withstand the storm as planned? Why or why not?

_____

_____

_____

_____

4. Compare your design to those of other groups. Suggest improvements to your design. What could you change to increase its strength?

_____

_____

_____

_____

CHECK YOUR HYPOTHESIS

Was your **hypothesis** supported by your data? Why or why not?

_____

_____

LAB 13.2   **DESIGN YOUR OWN**

CONCLUDE AND APPLY

1. Add your suggested improvements to the sketch of your design. Describe each of its functional parts. Explain how the improvements will increase the strength of the design.

   _____

   _____

   _____

   _____

   _____

   _____

2. Review your results and those of other groups. Which factors appear to affect the strength of the model homes?

   _____

   _____

   _____

   _____

3. In the tests, which held up best, solid or open walls? Why?

   _____

   _____

   _____

4. Infer why skyscrapers are designed to bend in the wind.

   _____

   _____

   _____

LAB ◆ **14.1** **INVESTIGATION**

Heat Absorption over Land and Water

If you have ever walked barefoot on the beach in the summer, you have noticed that the sand felt much hotter than the water. Why? Land and water absorb and release heat at different rates. In general, land heats up and cools down more quickly than does water. These differences affect coastal climates, making them cooler in the summer and warmer in the winter than areas farther inland.

PREPARATION

PROBLEM

How do soil and water compare in their abilities to absorb and release heat?

OBJECTIVES

- **Model** rates of heat absorption and heat release by land and water.
- **Measure** and **record** different rates of heat absorption and heat release in the air over land and water.
- **Analyze** the effects of heat absorption and heat release on climate.

MATERIALS

2 clear plastic boxes
4 alcohol-based thermometers
masking tape
water
soil
ring stand
overhead light with reflector
watch
colored pencils
metric ruler

SAFETY PRECAUTIONS

- Wear safety goggles, thermal mitts, and an apron during the lab procedure. Be careful when handling the hot overhead light. Do not let the light or its cord touch the water.
- If you break a thermometer, notify your teacher right away. Do not attempt to clean up broken glass.
- Avoid using mercury-based thermometers. Mercury is toxic.
- If possible, use a ground fault interruptor protected circuit (GFI) to prevent shock.

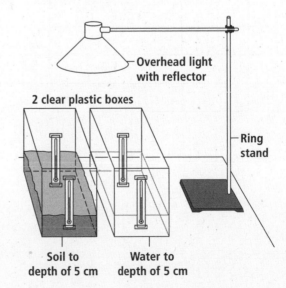

2 clear plastic boxes

Overhead light
with reflector

Ring
stand

Soil to
depth of 5 cm

Water to
depth of 5 cm

LAB **14.1** **INVESTIGATION**

PROCEDURE

1. Tape a thermometer inside a clear plastic box so that the thermometer bulb is 2 cm from the bottom of the box. Tape a second thermometer inside the same box so that the thermometer bulb is 8 cm from the bottom of the box. Label this box *A*.

2. Tape two thermometers inside the other plastic box, as in step 1. Label this box *B*.

3. Fill box *A* with room-temperature water to a depth of 5 cm. Fill box *B* with soil to a depth of 5 cm.

4. Attach the light to the ring stand and position the light above the boxes. Measure the temperatures of all the thermometers before turning on the light. Record your measurements in the table below.

5. Turn on the light. Adjust it to shed about the same light on each box. Record the temperature of each thermometer every minute for 10 minutes.

6. Turn off the light. Record the temperature of each thermometer every minute for 10 minutes.

DATA AND OBSERVATIONS

	Thermometer							
	Light on				Light off			
Minute	1	2	3	4	1	2	3	4
1								
2								
3								
4								
5								
6								
7								
8								
9								
10								

LAB 14.1 **INVESTIGATION**

DATA AND OBSERVATIONS, *continued*

LAB ◁ **14.1** ▷   **INVESTIGATION** 🔍

ANALYZE

1. Plot your data on the blank grid provided. Use a different colored pencil for the data from each thermometer. Describe any patterns you see in the data.

_____

_____

2. When the light was on, which heated up faster, the soil or the water? Compare heat-absorption rates for the air over the soil and the air over the water.

_____

_____

3. When the light was turned off, which substance lost heat faster, the soil or the water? Compare heat-loss rates for the air above the soil and the air above the water.

_____

_____

CONCLUDE AND APPLY

1. Summarize your results in terms of rates of heat absorption and heat release in land and water.

_____

_____

_____

2. How might rates of heat absorption and heat release affect air masses over land and water?

_____

_____

3. In January, City A at a latitude of 43° N has a temperature of 0°C. City B, at the same latitude, has a temperature of −5°C. Which city is probably closer to the coast? Explain your answer.

_____

_____

_____

4. Based on your results, explain why water in a swimming pool feels cool in the day and warm at night.

_____

_____

_____

LAB ◄ **14.2** ►

MAPPING

Use with
Section 14.2

Classifying Climates

A variety of factors influence the climate of an area, including latitude, topography, closeness to a lake or an ocean, availability of moisture, global wind patterns, and air masses. These various factors cause Earth's climates to range from scorching deserts to tropical rain forests to ice-covered polar regions. In general, climates are classified by natural vegetation and average monthly values for temperature and precipitation.

PREPARATION

PROBLEM
How do climates differ from one another?

OBJECTIVES
• **Interpret** climatic data on a world map.
• **Compare** and **contrast** different climates.

• **Analyze** the factors that make climates different.

MATERIALS
world map or globe

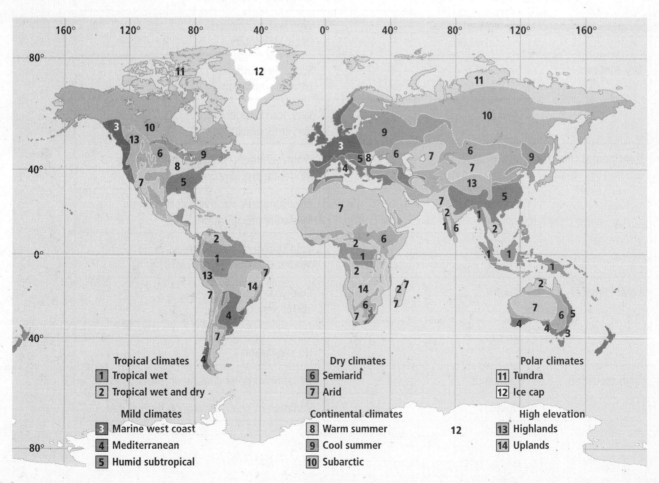

Tropical climates
1 Tropical wet
2 Tropical wet and dry

Mild climates
3 Marine west coast
4 Mediterranean
5 Humid subtropical

Dry climates
6 Semiarid
7 Arid

Continental climates
8 Warm summer
9 Cool summer
10 Subarctic

Polar climates
11 Tundra
12 Ice cap

High elevation
13 Highlands
14 Uplands

LAB ◁ **14.2** ▷   **MAPPING**

PROCEDURE

1. Carefully study the figure on the previous page. Note the latitudes and longitudes. Locate your area on the map to orient yourself. Compare the figure to a world map or globe that includes the names of countries.

You will need both maps to answer the questions on the next page.

2. Study the table below, which briefly describes the various types of climates included in the figure.

DATA AND OBSERVATIONS

Type of Climate	Description	Defining Characteristics
Tropical climates	Tropical wet	High temperatures year-round High rates of precipitation
	Tropical wet and dry	High temperatures year-round Wet summers, dry winters
Mild climates	Marine west coast	Cool summers, mild winters Abundant precipitation
	Mediterranean	Warm summers, mild winters Moderate precipitation
	Humid subtropical	Wet, warm summers Dry, cool winters
Dry climates	Semiarid	Scarce vegetation Little precipitation
	Arid	Very scarce vegetation Very little precipitation
Continental climates	Warm summer	Warm summers, relatively cold winters Moderate precipitation
	Cool summer	Cool summers, relatively cold winters Moderate precipitation
	Subarctic	Cold summers, cold winters Moderate precipitation
Polar climates	Tundra	Cold year-round Scarce vegetation
	Ice cap	Very cold year-round No vegetation
High elevation	Highlands	Variation of polar climate on mountains
	Uplands	Variation of polar climate on high plateaus

LAB **14.2**   **MAPPING**

ANALYZE

1. What is the climate of your area? Use the table to describe your climate.

   _____

   _____

2. What country is located at 25°N, 20°E? Describe the climate of this country.

   _____

   _____

3. In terms of latitude, where are most tropical wet climates located?

   _____

   _____

4. Study the locations of marine west coast and mediterranean climates. What factor appears to have the most influence on these climates?

   _____

5. Compare and contrast the climates of the west and east coasts of Nicaragua.

   _____

   _____

   _____

   _____

   _____

6. Clashes between air masses can cause extreme variations in temperature. In the United States, which climate types appear to be most affected by air masses? Explain your answer.

   _____

   _____

   _____

LAB 14.2   **MAPPING**

CONCLUDE AND APPLY

1. How do climates differ on either side of the Rocky Mountains in the northwest United States? What can you infer about the influence of the mountains on climate?

_____

_____

_____

_____

_____

_____

2. Which part of Australia would be best suited for growing crops that need plenty of moisture and mild temperatures year-round? Why?

_____

_____

_____

3. South America and Africa do not extend to the poles, yet parts of these continents experience polarlike climates. Why?

_____

_____

_____

LAB **15.1** MAPPING

**Use with
Section 15.2**

Ocean Surface Temperatures

Ocean water has distinct chemical and physical properties, such as level of salinity, temperature, and the ability to absorb light. Because the oceans constantly intermix, these properties can vary from day to day and from place to place. Scientists use satellite data to track changes in some of these properties. Ocean surface temperatures, for example, can be determined by using satellite imagery that detects differences in thermal energy. These data are then compiled into maps.

PREPARATION

PROBLEM
How do ocean surface temperatures vary from place to place?

MATERIALS
globe

OBJECTIVES
- **Interpret** a world map of ocean surface temperatures.
- **Compare** the surface temperatures of different oceans.
- **Analyze** why ocean surface temperatures vary.

PROCEDURE

1. Study the map, which shows ocean surface temperatures in October 2000. Compare it to a globe.

2. Use the globe to label the oceans and the continents on the map. Add latitude and longitude coordinates to the map. Also label north, south, east, and west on the map.

LAB **15.1**   **MAPPING**

DATA AND OBSERVATIONS

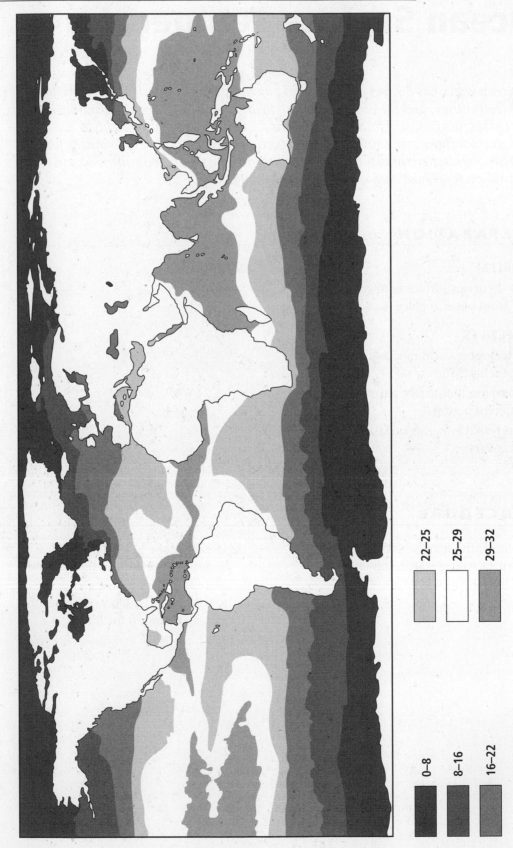

LAB **15.1** MAPPING

ANALYZE

1. What is the range of ocean surface temperatures shown in the scale on the map?

   _____

   _____

2. Look for and describe patterns on the map. For example, which surface temperature or range of temperatures appears to be most common?

   _____

   _____

   _____

   _____

   _____

   _____

3. What is the surface temperature of the ocean nearest to the place you live? Convert this temperature to the Fahrenheit scale.

   _____

   _____

4. Describe how ocean surface temperatures change from the northern Pacific Ocean southward to Antarctica.

   _____

   _____

   _____

   _____

   _____

   _____

LAB **15.1** **MAPPING**

CONCLUDE AND APPLY

1. Where are the coldest surface temperatures found? Where are the warmest found? What accounts for these differences in temperature?

_____

_____

_____

_____

_____

2. Global warming is an increase in global temperatures caused by increases in certain atmospheric gases. How might scientists use maps such as the one in this lab to analyze global warming? What might be some other uses of this map?

_____

_____

_____

_____

_____

_____

3. How might this map change if the satellite data were gathered in February? In July?

_____

_____

_____

_____

LAB 15.2 INVESTIGATION

Use with Section 15.3

Making Waves

As wind blows across the surface of the ocean, friction causes the water to move with the wind. If conditions are right, water begins to pile up and forms a wave. Ocean waves vary greatly in height. Those that reach shallow water and break on shore may be less than 1 m high. In the open ocean, waves can reach towering heights of 30 m. A monster wave is usually caused by a powerful storm.

PREPARATION

PROBLEM
What factors affect the heights of waves?

OBJECTIVES
- **Model** the movement of waves.
- **Measure** and **record** differences in wave heights.
- **Infer** what factors affect the heights of waves.

MATERIALS
electric fan with variable speed and a grounded/polarized plug
overhead light with reflector
ring stand
white paper
timer
clear, shallow, rectangular container
water
metric ruler

SAFETY PRECAUTIONS

- If you are using any electrical outlets near water, be sure there is GFI (ground fault interruptor) protection.
- Wear safety goggles during the lab procedure.
- Be careful when handling the overhead light and the fan. The light could get very hot. Do not let the light, the fan, or their cords touch the water.
- Do not stick your fingers or other objects in the fan blades.
- Wipe up spills immediately to help prevent slipping or falling.

PROCEDURE

1. Lay a large sheet of white paper on a flat surface, then place the container on the paper. Position a ring stand to the side of the container. Clamp a light on the ring stand so that the light shines directly into the container.

2. Fill the container nearly to the top with water. Place a fan at one end of the container. Turn it on low.

3. After 3 minutes, measure the heights of the waves created by the fan. Record your measurements in the table provided.

 **LAB 15.2**

PROCEDURE, *continued*

4. Keep the fan on low and carefully observe the shadows of the waves on the white paper. Record your observations.

5. After 5 minutes, measure and record the heights of the waves again.

6. Repeat steps 3–5 with the fan on medium speed and on high speed.

7. Turn off the fan. Observe and record what happens to the water.

DATA AND OBSERVATIONS

Fan Speed	Wavelength	Observations
Low, 3 minutes		
Low, 5 minutes		
Medium, 3 minutes		
Medium, 5 minutes		
High, 3 minutes		
High, 5 minutes		

LAB **15.2** **INVESTIGATION**

ANALYZE

1. Compare the heights of the waves when the fan was on low, medium, and high speeds.

   _____

   _____

   _____

   _____

   _____

2. Did the heights of the waves change with time? Explain your answer.

   _____

   _____

   _____

   _____

   _____

3. Describe how the shadows of the waves changed when the speed of the fan changed.

   _____

   _____

   _____

   _____

   _____

4. Describe the movement of the water when the fan was turned off.

   _____

   _____

   _____

   _____

LAB ◁ **15.2** ▷ **INVESTIGATION**

CONCLUDE AND APPLY

1. Based on your results, what factors influence wave height?

_____

_____

_____

2. What might happen to wave height if you repeated the experiment using a much longer container? What might happen if the container was deeper?

_____

_____

_____

_____

_____

_____

3. The heights of ocean waves vary greatly. Describe some conditions that might generate both high and low ocean waves.

_____

_____

_____

_____

_____

Use with
Section 16.1

Changes in Sea Level

Sea level continually changes in response to numerous factors, including glacial melting, tectonic forces, and climatic changes. Currently, sea level is rising at a rate of 1.5 to 3.9 mm/year. Studies by the U.S. Environmental Protection Agency and other organizations indicate that global warming may be linked to increases in sea level. Global warming is a worldwide rise in surface temperatures. Scientists hypothesize that global warming is caused by human activities, such as the burning of fossil fuels. Even a small rise in global temperatures can melt glaciers and make seawater expand, both of which increase the volume of water in the oceans.

PREPARATION

PROBLEM

How have coastlines and sea level changed during geologic time?

MATERIALS

ruler
string

OBJECTIVES

- **Observe** and **measure** changes in coastlines.
- **Describe** changes in sea level over geologic time.
- **Predict** the impact of rising sea level on coastal regions.

PROCEDURE

1. Take a few minutes to study the sea-level map on the next page. Note the map scale. You will use the map and the map scale to answer the questions in this lab.

2. Use the string to measure distances along the coast or between two points that are not in a straight line. For example, you can lay the string along the coast so that it follows the outline of the coast, then measure the distance by laying the string along a ruler.

LAB ◆ **16.1** **MAPPING**

DATA AND OBSERVATIONS

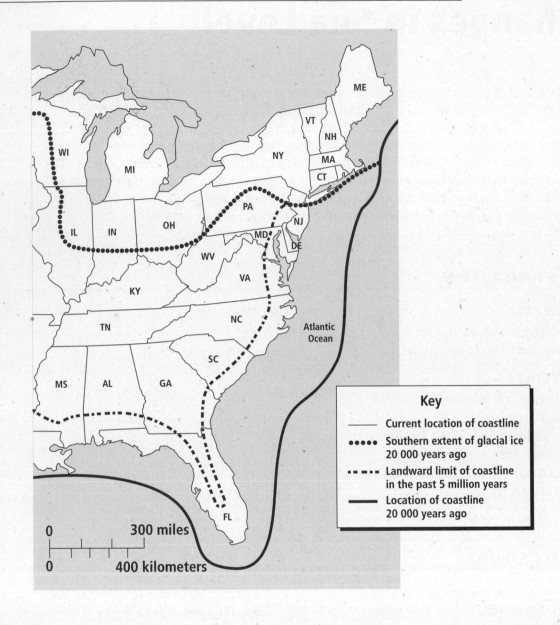

Key

— Current location of coastline

••••• Southern extent of glacial ice 20 000 years ago

– – – Landward limit of coastline in the past 5 million years

— Location of coastline 20 000 years ago

LAB **16.1**   **MAPPING**

ANALYZE

1. How does the landward limit of the coastline 5 million years ago compare to its current location? How does it compare to the location of the coastline 20 000 years ago?

_____

_____

_____

_____

2. Locate South Carolina on the map. How far did its coastline extend into the Atlantic Ocean 20 000 years ago? How far inland was its coastline 5 million years ago?

_____

_____

_____

_____

3. Measure and record the entire length of the current coastline and its length 20 000 years ago. Describe how it has changed. What caused these changes?

_____

_____

_____

_____

_____

4. Use the map to describe how sea level has changed in the last 20 000 years. Why do you think these changes occurred?

_____

_____

_____

_____

LAB 16.1   **MAPPING**

CONCLUDE AND APPLY

1. The last ice age peaked roughly 10 000 years ago. Since then, sea level has risen approximately 130 m. Describe the effect of small rises in sea level on coastal areas.

_____

_____

_____

2. The mass of huge glaciers exerts pressure on underlying land and causes it to sink. When these glaciers retreat, the land that they covered often rises, or rebounds. Where on the map would you expect glacial rebound to be occurring? How might this rebound affect the levels of seas and other large bodies of water?

_____

_____

_____

3. Global sea level could rise by 30 cm within the next 70 years. Predict which areas on the map would be affected most. Explain your answer.

_____

_____

_____

4. Discuss the potential impacts of rising sea level on low-lying coastal areas.

_____

_____

_____

_____

LAB 16.2   **INVESTIGATION**

Use with
Section 16.2

Observing Brine Shrimp

Oceans cover 70 percent of Earth, and these vast waters support an amazing array of marine life. Marine organisms range from massive whales to microscopic plankton, but all marine organisms must be able to survive in seawater. In this lab, you will examine how salinity affects aquatic organisms called brine shrimp. Brine shrimp are tiny crustaceans that have hard exoskeletons, jointed legs, and antennae. They live in salty lakes and ponds.

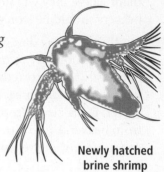

**Newly hatched
brine shrimp**

PREPARATION

PROBLEM
Under what conditions do brine shrimp hatch and thrive?

OBJECTIVES
- **Culture** brine shrimp.
- **Observe** and **record** data about the structure and behavior of a crustacean.
- **Analyze** the effects of different salt concentrations on an aquatic organism.

MATERIALS
500-mL beakers (4)
glass marker
100-mL graduated cylinder
room-temperature water
alcohol-based thermometer
75 g noniodized salt
balance
4 stirring rods

wooden stick
brine-shrimp eggs
plastic wrap
4 droppers
4 petri dishes
microscope

SAFETY PRECAUTIONS

- If you break a thermometer or beaker, notify your teacher right away. Do not attempt to clean up broken glass.
- Avoid using mercury-based thermometers. Mercury is toxic.
- Wipe up any spills immediately.
- Wear safety goggles during the lab procedure. Salt water can irritate your eyes.

PROCEDURE

1. Use a glass marker to label four beakers A, B, C, and D. Measure 500 mL of room-temperature water into each beaker. Use a marker to label the petri dishes similarly.

2. Add 5 g of noniodized salt to beaker B. Add 20 g of salt to beaker C. Add 50 g of salt to beaker D. Do not add salt to beaker A.

3. Use the rounded end of a wooden stick to transfer some brine-shrimp eggs to beaker A. Mix the solution with a stirrer. The eggs are tiny and difficult to count, but try to transfer roughly the same number of eggs into beakers B and C. Use a clean, dry stirrer each time.

4. Record the temperature of the water in each beaker in a data table.

LAB ◁ **16.2** ▷   **INVESTIGATION**

PROCEDURE, *continued*

5. Cover the beakers with plastic wrap and store them where they will not be disturbed. All beakers should be stored at around 21°C.

6. On the following day, uncover the beakers and use a thermometer to measure water temperatures. Record these temperatures in your data table. Stir each beaker with a clean stirrer. Use a clean dropper to place a few drops from beaker A into a petri dish. Repeat this procedure for the remaining solutions.

7. Observe the petri dishes under a microscope. Count the number of hatched brine shrimp in each dish. Record your observations in Table 1.

8. Repeat steps 5 through 7 for 3 more days.

DATA AND OBSERVATIONS

Table

Beaker	Day 1	Temperature	Day 2	Temperature	Day 3	Temperature	Day 4	Temperature
				Number of Hatched Brine Shrimp				
A								
B								
C								
D								

LAB ◅ **16.2** ▻   **INVESTIGATION**

ANALYZE

1. How long did it take the brine shrimp to hatch? Did they hatch in all the beakers?

_____

_____

_____

2. Which beaker had the most hatched brine shrimp? Which beaker had the least?

_____

_____

_____

3. Calculate the percentage of salt in each solution, using grams per milliliter.

_____

_____

_____

4. Describe the structure and color of the hatched brine shrimp. Be specific. Draw one brine shrimp below Table 1.

_____

_____

_____

_____

5. Describe the behavior of the brine shrimp. For example, how do they move about in the water?

_____

_____

_____

_____

_____

_____

LAB ◁ **16.2** ▷   **INVESTIGATION**

CONCLUDE AND APPLY

1. Under which conditions did brine shrimp thrive best?

_____

_____

_____

2. How might changes in salinity affect the brine shrimp?

_____

_____

_____

_____

_____

3. Like many marine organisms, brine shrimp respond to light. Based on this information, what can you conclude about their natural habitat?

_____

_____

_____

_____

_____

LAB **17.1**   **DESIGN YOUR OWN**

Use with
Section 17.2

Magnetism and Ocean Ridges

Evidence of seafloor spreading includes magnetic measurements of the basaltic rocks at the bottom of the ocean. Earth's magnetic field at different places along the seafloor is stronger than Earth's current field, and some places have a weaker than normal field. These variations indicate that Earth's magnetic field reverses periodically over geologic time.

The plausibility of this interpretation can be demonstrated with a model. You can put iron filings in a test tube, set a magnet near the end of the table, and tap it. The individual filings become magnetized and line up, effectively becoming a collection of small bar magnets that all face in the same direction. The magnetized iron filings model iron-bearing minerals in molten material that wells up along a divergent plate boundary. As the molten material cools, it preserves the magnetic orientation at the time of crystallization. Magnetic fluctuations on the seafloor form symmetrical patterns in relation to an ocean ridge (Figure 1).

PREPARATION

PROBLEM

How are patterns of magnetic-field strength in seafloor rocks related to changes in the polarity of Earth's magnetic field?

OBJECTIVES

- **Investigate** the mechanism of magnetization.
- **Model** how magnetic patterns preserved in seafloor rocks arise.

HYPOTHESIS

With your group, write a description of how patterns of magnetic fluctuations arise on the seafloor.

_____

_____

_____

POSSIBLE MATERIALS

test tube
test-tube stopper
iron filings
bar magnet
small plotting compass
meterstick

SAFETY PRECAUTIONS

Use caution when handling iron filings. Wear safety goggles to help keep filings away from your eyes.

Figure 1

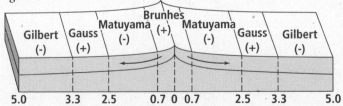

Age (millions of years)

LAB 17.1

PLAN THE EXPERIMENT

Consider the possible materials and design an experiment that supports the theory that patterns observed in seafloor rocks indicate reversals in Earth's magnetic field. Seafloor measurements do not indicate that the field is currently reversed; rather, they show that the field is weaker or stronger than the average. So, you need to confirm that a weak magnetic field (like that of iron-bearing minerals in rocks as represented by the iron filings), in the presence of a strong magnetic field (like that of Earth as represented by the bar magnet), can produce a combined field that is stronger or weaker than the average magnetic field is alone. What can you use to discover the turning point, where the field's effect changes? List the materials you will need and outline your experiment in Data and Observations. You may also choose to draw a diagram as part of your outline. Have your teacher approve your plan before you start.

DATA AND OBSERVATIONS

OUTLINE OR DIAGRAM

_____

_____

_____

_____

_____

_____

_____

_____

_____

_____

LAB **17.1**   **DESIGN YOUR OWN**

ANALYZE

1. Consider the magnetic field strength at the following four locations: the surface of Earth, the surface of a bar magnet, the surface of a test tube containing magnetized iron filings, and the seafloor. List these four surfaces from strongest to weakest.

_____

_____

_____

2. How far from the bar magnet does its field equal Earth's magnetic field? How could you tell?

_____

_____

_____

3. If you were to move the compass a few extra centimeters away from the magnet, do you think the strength of the combined magnetic field at that point would be greater than, less than, or equal to the strength of Earth's magnetic field? Why?

_____

_____

_____

4. Do the magnetized iron filings have a stronger or weaker magnetic field than the bar magnet?

_____

_____

CHECK YOUR HYPOTHESIS

Was your **hypothesis** supported by your data? Why or why not?

_____

_____

_____

LAB **17.1** **DESIGN YOUR OWN**

CONCLUDE AND APPLY

1. Measurements of seafloor magnetic fields show regions of stronger-than-normal and weaker-than-normal fields. The measurements do not actually indicate reversed fields. Explain how the measurements provide evidence that Earth has experienced periods of reversed magnetic polarity.

_____

_____

_____

_____

2. Look at Figure 1. Why is there magnetic symmetry in relation to an ocean ridge?

_____

_____

_____

3. If the ocean ridge ran east-west instead of north-south, would you expect to see patterns of magnetic symmetry in the seafloor rock? Explain your answer.

_____

_____

_____

_____

4. When might the magnetic field of Earth reverse? Explain your answer.

_____

_____

_____

_____

INVESTIGATION

Use with
Section 17.3

Earthquakes and Subduction Zones

T he density of the rock that makes up a subducting
plate is one of the factors that determines how the
plate behaves. The greater the density, the faster the plate
subducts into the mantle and the steeper the angle of
subduction. Older crust is cooler and therefore denser than
younger crust, so it subducts faster and at a steeper angle along a subduction zone.

Figure 1

Most earthquakes occur at tectonic plate boundaries. An earthquake can be classified by the depth
of its focus. Deep-focus earthquakes have foci at more than 300 km, shallow-focus earthquakes have
a focus at less than 70 km, and intermediate-focus earthquakes have foci between 70 km and 300 km.

PREPARATION

OBJECTIVES

- **State** a hypothesis about the relative
  ages of the crust at two convergent
  boundaries.
- **Use** earthquake data to **construct** pro-
  files of two convergent boundaries.
- **Compare** the behavior of two subducting plates.

HYPOTHESIS

Consider Figure 1. The East Pacific Rise
is an ocean ridge, running north-south
at about 110°W, where the Pacific Plate
meets the Nazca Plate. Material from
this divergent boundary flows westward
across the Pacific Plate or eastward
across the Nazca Plate. The west-
flowing material runs into the
Australian Plate at the Tonga Trench,

which is north of New Zealand at about
175°W. East-flowing material meets the
South American Plate at the Peru-Chile
Trench, at about 65°W. Assume that the
seafloor spreads at the same rate both
west and east of the East Pacific Rise.
Form a hypothesis about the relative
ages of the East Pacific Rise material at
the two convergent boundaries: the
Tonga Trench and the Peru-Chile Trench.

_____

_____

_____

_____

MATERIALS
calculator

PROCEDURE

1. Table 1 shows earthquake data
   from the region associated with the
   Peru-Chile Trench. Plot these data
   on a graph, using a dot to
   represent each data point.

2. Plot the earthquake data from the
   region associated with the Tonga
   Trench on a second graph.

3. Draw a best-fit line for the Peru-
   Chile Trench data. A best-fit line is
   a smooth line that shows the trend
   of the data; the line does not have
   to pass through the data points.

4. Draw a best-fit line for the Tonga
   Trench data.

LAB **17.2**

DATA AND OBSERVATIONS

Peru-Chile Trench		Tonga Trench	
Longitude (°W)	Focus depth (km)	Longitude (°W)	Focus depth (km)
61.7	540	173.8	35
62.3	480	173.8	50
63.8	345	173.8	60
65.2	285	173.9	60
65.5	290	174.1	30
66.2	230	174.6	40
66.3	215	174.7	35
66.4	235	174.8	35
66.5	220	174.9	40
66.7	210	174.9	50
66.7	200	175.1	40
66.9	175	175.4	250
67.1	230	175.7	205
67.3	185	175.7	260
67.5	180	175.8	115
67.5	170	175.9	190
67.7	120	176.0	160
67.9	140	176.0	220
68.1	145	176.2	270
68.1	130	176.8	340
68.2	160	177.0	380
68.3	130	177.0	350
68.3	110	177.4	420
68.4	120	177.7	560
68.5	140	177.7	580
68.6	180	177.7	465
68.6	125	177.8	460
69.1	95	177.9	565
69.2	35	178.0	520
69.3	60	178.1	510
69.5	75	178.2	595
69.7	50	178.2	550
69.8	30	178.3	540
69.8	55	178.5	505
70.8	35	178.6	615
		178.7	600
		178.8	590
		178.8	580
		179.1	675
		179.2	670

LAB **17.2**   **INVESTIGATION**

DATA AND OBSERVATIONS, *continued*

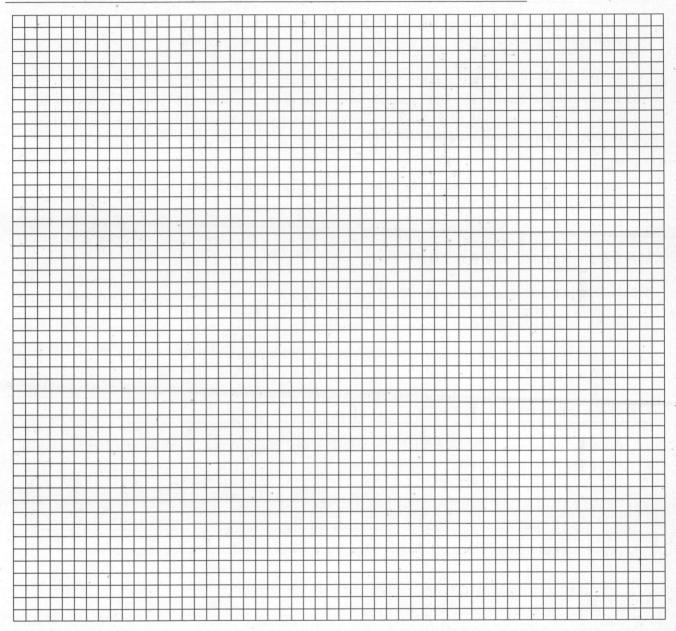

ANALYZE

1. How far is the Tonga Trench from the East Pacific Rise? Note that one degree longitude equals about 100 km. If the seafloor spreads at 3 cm/year, how long would it take material on the plate to travel this distance?

_____

2. What is the depth of the deepest earthquake in the Tonga data set? Estimate the rate of descent of the East Pacific Rise material at the Tonga Trench in centimeters per year.

_____

LAB  **17.2**

ANALYZE, *continued*

3. Estimate the rate of descent of East Pacific Rise material into the Peru-Chile Trench in centimeters per year.

_____

_____

4. The best-fit line on the Peru-Chile graph is an estimate of the location of the boundary between the Nazca Plate and the South American Plate. Indicate on the graph which plate is which. Add an arrow to show the direction of motion of the Nazca Plate.

5. The best-fit line on the Tonga graph is an estimate of the location of the boundary between the Pacific Plate and the Australian Plate. Indicate on the graph which plate is which. Add an arrow to show the direction of motion of the Pacific Plate.

CHECK YOUR HYPOTHESIS

Was your **hypothesis** supported by your data? Why or why not?

_____

_____

CONCLUDE AND APPLY

1. Compare your two graphs. Which has the steeper profile? Which do you think has the denser material? The older material? Explain your answer.

_____

_____

2. Summarize your observations, including a statement about the validity of your hypothesis.

_____

_____

_____

_____

_____

_____

LAB 18.1   **DESIGN YOUR OWN**

Use with
Section 18.2

Modeling a Lava Flow

*L*ava flows can form a complicated variety of structures. Levees can form on the outer part of a lava-flow channel, and ridges sometimes develop inside the flow. Shear zones can occur where one part of the flow moves faster than another. Often the easiest way to explore complicated physical systems such as lava flows is to make a model. You can use a model to study how slope and the position in a flow affect the dimensions of the flow.

PREPARATION

PROBLEM

How is a lava flow affected by changes in the conditions that produce it?

OBJECTIVES

- **Model** the geologic processes associated with lava flows.
- **Collect** data on the structure of a model lava flow.

HYPOTHESIS

The slope over which a lava flow travels and lava viscosity are two parameters that determine how a lava flow behaves. What effects do these two parameters have on the width, depth, length, and speed of a flow? How do these two parameters affect structures in a flow? Write a hypothesis about what your experiment might show about flow characteristics.

_____

_____

_____

_____

POSSIBLE MATERIALS

dry cake mix
water
bucket with pouring spout
wire whisk
coffee tin with 2-cm circular
   hole in the bottom
large spatula
wooden board, 1 m × 2 m
paper
plastic wrap
wooden shims and wedges
protractor
plumb line
stopwatch
metric ruler
toothpicks
masking tape

SAFETY PRECAUTIONS

- Wear safety goggles and an apron during the lab procedure.
- Immediately clean up floor spills of cake mix; they can be slippery.
- Handle the coffee tin carefully. The hole in the bottom may have sharp edges that can cut your skin.
- Follow your teacher's instructions to dispose of the cake mix at the end of the lab.

LAB  **18.1**   **DESIGN YOUR OWN**

PLAN THE EXPERIMENT

Work with four other students. Review the list
of possible materials. Design an experiment to
simulate the evolution of lava flows on the side
of a volcano. You should consider the slope of
the incline, pouring rate, and viscosity of the
lava. What cases will you consider? How will you
measure the variables? What data will you collect?
Set up a table or tables in which to record the data.
Plan to draw diagrams, including cross sections, of
the flow features. Outline your plan and have your
teacher approve it before you begin the
experiment.

_____

_____

_____

_____

_____

_____

_____

_____

_____

_____

_____

_____

_____

_____

_____

_____

_____

_____

DATA AND OBSERVATIONS

TABLES AND DIAGRAMS

LAB ◀ **18.1** ▶   **DESIGN YOUR OWN**

DATA AND OBSERVATIONS, *continued*

ANALYZE

1. Use your diagrams to describe how the flows evolved.

_____

_____

_____

_____

LAB **18.1**   **DESIGN YOUR OWN**

ANALYZE, *continued*

2. Describe the nature and locations of any flow channels, levees, ridges, and shear zones that you observed.

_____

_____

_____

_____

3. Summarize your results with respect to the slope of the incline, viscosity, and pouring rate.

_____

_____

_____

CHECK YOUR HYPOTHESIS

Was your **hypothesis** supported by your data? Why or why not?

_____

_____

CONCLUDE AND APPLY

1. How well did your model represent a real lava flow? Describe the model properties that are identical to those of a lava flow, similar but on a different scale, and not similar at all.

_____

_____

_____

_____

2. Shield volcanoes have gently sloping sides, whereas cinder-cone volcanoes and composite volcanoes usually have steeper sides. Use the results of your experiment to predict how lava flows differ for these types of volcanoes.

_____

_____

_____

LAB ◀ **18.2** **INVESTIGATION**

Analyzing Volcanic-Disaster Risk

On May 18, 1980, an earthquake shook Mount St. Helens. A bulge on the side of the mountain and the area surrounding it slid away in a gigantic avalanche, releasing pressure and triggering a major pumice and ash eruption of the volcano. Debris filled 62 km^2 of a valley; a lateral blast damaged 650 km^2 of recreation, timber, and private lands; and volcanic mud flows deposited an estimated 0.15 km^3 of material in the nearby river. Nearly five dozen people died in the eruption. There was over $1 billion in damage.

PREPARATION

PROBLEM

What is the probability that a volcano will erupt in any given year? What does that imply for the cost of insuring people against volcanic disasters?

MATERIALS

Tables 1 and 2
calculator

OBJECTIVES

- **Assess** the probability of a volcanic disaster.
- **Investigate** the feasibility of an insurance policy against volcanic disaster.

PROCEDURE

Mount St. Helens is a volcano in the Cascade Range, which extends from California to British Columbia. Table 2 contains data for the eruption histories of the Cascade Range volcanoes. With these data, you could estimate the annual probability that a particular volcano will erupt.

$$\text{annual probability} = \frac{\text{number of eruptions}}{\text{years}}$$

For example, based on the Holocene data, the probability that Lassen Peak will erupt in any given year is (3 eruptions)/(10 000 years),

or 3/10 000. From Table 1, you can see that this value lies between the annual probabilities that an individual human will die by homicide or die of AIDS.

Only huge eruptions left records before the Holocene. Smaller eruptions in the Pleistocene are poorly documented in the rock record. To calculate eruption probabilities for which there are no Pleistocene data, use the 10 000-year Holocene baseline.

Use the formula above and the data in Table 2 to answer the questions in Analyze.

LAB 18.2   INVESTIGATION

DATA AND OBSERVATIONS

Table 1

Event	Annual Probability per Person
Experience car theft	1/100
Experience house fire	1/200
Die from heart disease	1/280
Die of cancer	1/500
Die in a car wreck	1/6000
Die by homicide	1/10 000
Die of AIDS	1/11 000
Die of tuberculosis	1/200 000
Win a state lottery	1/1 million
Die from lightning	1/1.4 million
Die from a flood or tornado	1/2 million
Die in a hurricane	1/6 million
Die in a commercial plane crash	1/1 million to 1/10 million

Table 2

Feature	State	Late Pleistocene Huge	Holocene Huge	Large	Medium	Small
Mount Baker	WA				1	3
Glacier Peak	WA			2		7
Mount Rainier	WA	1			1	10
Mount St. Helens	WA				7	
Mount Adams	WA					4
Mount Hood	OR					3
Mount Jefferson	OR	1				
Three Sisters	OR			2		2
Newberry Caldera	OR			1	3	
Crater Lake	OR		1	2		
Medicine Lake	CA				8	8
Mount Shasta	CA	1		2		10
Lassen Peak	CA	4		1		2
Total		7	1	10	20	49

Note: Huge eruptions = >10 km^3 of ejecta; large = 1–10 km^3; medium = 0.1–1 km^3; small = <0.1 km^3.
The periods of the eruptions are late Pleistocene (10 000 years to 100 000 years ago) or Holocene (<10 000 years ago).

LAB **18.2**   **INVESTIGATION**

ANALYZE

1. Based on the Holocene data, what is the probability that Medicine Lake will erupt this year?

_____

2. What is the probability that Medicine Lake will have a medium-size eruption this year?

_____

3. What is the annual probability that Mount Jefferson will erupt?

_____

4. During the last 10 000 years, how many eruptions have occurred in the Cascade Range?

_____

5. What is the annual probability that one of the Cascade Range volcanoes will erupt?

_____

6. What is the annual probability that a small eruption will occur in the Cascade Range? Multiply this by $100 million to obtain the contribution of small eruptions to the annual cost of Cascade volcanic disasters.

_____

7. What is the annual probability that a medium-size eruption will occur in the Cascade Range? Multiply this by $1 billion to obtain the contribution of medium-size eruptions to the annual cost of Cascade volcanic disasters.

_____

8. Calculate the annual probability for large Cascade eruptions. If a large Cascade eruption costs $10 billion, what is the contribution of large eruptions to the annual cost of Cascade volcanic disasters?

_____

9. Using both Pleistocene and Holocene data, what is the annual probability that a huge eruption will occur? Calculate the contribution of huge eruptions to the annual cost of Cascade volcanic disasters if the cost of a huge eruption is $100 billion.

_____

10. What is the total cost per year of Cascade eruptions? If inflation and land development drove the cost per eruption up by a factor of 100, what would the annual cost be?

_____

LAB ◄ **18.2** ►   🔍 **INVESTIGATION**

CONCLUDE AND APPLY

1. Think of two reasons why your estimates of the probabilities might predict three times fewer volcanic eruptions than the actual number that will occur during the twenty-first century.

   _____

   _____

   _____

   _____

   _____

2. If there were three times as many volcanoes as predicted, what would the total annual cost be? Use the value that you obtained for question 10. If this cost is distributed evenly among 2 million policyholders, what would the insurance premium be?

   _____

   _____

3. Suppose that 10 percent of the inhabitants in the Cascade Range owned 90 percent of the property, and that the cost of the insurance premium would prohibit the other 90 percent from becoming policyholders. Divide 90 percent of the cost by 10 percent of 2 million people to find out what the annual insurance premium would be.

   _____

   _____

4. Explain why insurance companies rarely insure against volcanic-disaster damage, even though volcanic disasters do not occur very often.

   _____

   _____

   _____

   _____

   _____

LAB ◆ **19.1** **INVESTIGATION**

Use with
Section 19.3

Predicting Earthquakes

The United States Geological Survey (USGS) compiles and posts on the Internet a list of recent earthquakes. This list is updated every 5 minutes. In a typical month, the list may include 300–400 earthquakes that occurred around the world. The list includes each earthquake's location, magnitude, and depth. These data can be mapped and used to predict where earthquakes will occur next.

PREPARATION

PROBLEM

Where are earthquakes most likely to occur next ?

MATERIALS

computer with Internet access
fine-point pens, 6 different colors

OBJECTIVES

- **Analyze** the locations, magnitudes, and depths of recent earthquakes.
- **Predict** where earthquakes are most likely to occur in the next few weeks.

PROCEDURE

1. Visit sites listed on the Glencoe Science Web Site to retrieve and print out the USGS table of earthquake data.

2. Identify the columns for the time of the earthquake and its latitude, longitude, depth, and magnitude. Use the body-wave magnitude.

3. Decide on a scheme of colors and symbols to map earthquake locations. A symbol's color should indicate the magnitude of the earthquake, and its shape should identify the depth. Define your scheme in the two keys in Figure 1.

4. Identify the data for the most recent earthquake in the USGS table. Use a sharp pencil to plot the position of the earthquake on the world map in Figure 1.

5. Ask your teacher to verify that you have correctly plotted the earthquake's location.

6. Use a fine-point pen and follow your symbol scheme to mark the point.

7. Plot each earthquake on the list or as many as your teacher specifies. This task will go more quickly if one person reads the data aloud while another person plots the point.

8. Locate an area 5° × 5° on the world map where you expect an earthquake to occur in the next few weeks.

9. Use your data to complete columns 1 and 2 of Table 1.

LAB 19.1   **INVESTIGATION**

DATA AND OBSERVATIONS

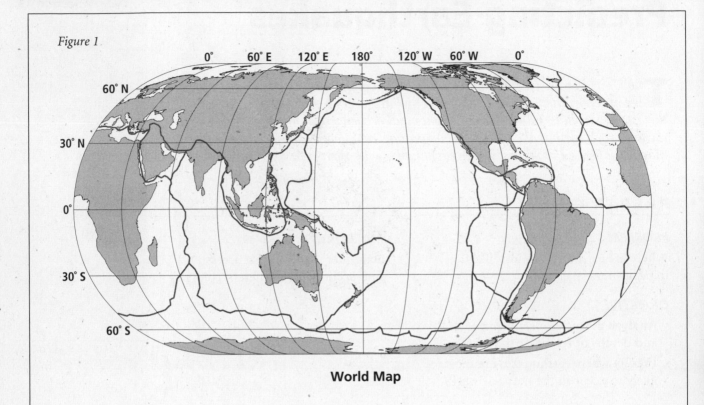

Figure 1

World Map

Magnitude	Color
<4	
4 – 4.9	
5 – 5.9	
6 – 6.9	
7 - 7.9	
>8	
no data	

Depth (km)	Symbol
0 – 50	
51 – 100	
101 – 150	
151 – 200	
201 – 250	
>250	
no data	

LAB **19.1**   **INVESTIGATION**

DATA AND OBSERVATIONS, *continued*

Table 1

	Actual Earthquakes		Predicted Earthquakes in the 5° × 5° Area	
	Number of Earthquakes	**Percentage of Earthquakes**	**Predicted Earthquakes**	**Actual Earthquakes**
Richter Magnitude				
<4				
4–4.9				
5–5.9				
6–6.9				
7–7.9				
>8				
No data				
Depth (km)				
0–50				
51–100				
101–150				
151–200				
>200				
No data				

LAB **19.1**   **INVESTIGATION**

ANALYZE

1. Find the time interval, to 0. 1 days, between the least recent and most recent earthquakes for the data you plotted. Calculate how many earthquakes occurred per day during this interval. This is the earthquake rate.

_____

2. What is the earthquake rate in your 5° × 5° area on the map?

_____

3. Predict how many earthquakes you expect to occur in your 5° × 5° area during the next week. Give reasons for your prediction.

_____

_____

_____

4. Use the data in Table 1 to help you predict for the next few weeks the number of earthquakes in each magnitude range and each depth range in your 5° × 5° area. Record your predictions in column 3 of Table 1.

CONCLUDE AND APPLY

1. Earth's major tectonic plate boundaries are shown in Figure 1. What can you hypothesize about the distribution of earthquake locations?

_____

_____

2. A week later, get the past week's data from the web site. Use the new data to complete column 4 in Table 1 for your 5° × 5° area . How well do your predictions match the actual events of the past week? Explain your answer.

_____

_____

_____

LAB 19.2 — **DESIGN YOUR OWN**

Earthquake News Report

Many scientific concepts can be described in depth to an audience with no in-depth knowledge of the subject. The key is to consider at the outset the target idea that you want to convey. A scientific idea usually builds off a few other ideas. Once those lower-level ideas are understood, then the target idea is easily accessible. If these lower-level ideas can be presented by analogy to common experience, they are contact points. If a lower-level idea is not a contact point, it can be treated as a target idea that depends on even lower-level ideas, and so on. Most target ideas have a reasonably manageable set of contact points. Conveying the target idea requires establishing contact points with the audience and weaving the points together to construct the target idea.

PREPARATION

PROBLEM

How can you convey technical information to a popular audience?

OBJECTIVES

- **Address** the issues involved in communicating technical ideas to a popular audience.
- **Make** a team presentation about earthquakes.
- **Critique** technical presentations in a constructive way.

HYPOTHESIS

What are the three most important things that a scientist should consider when communicating technical ideas to an audience that may not have a technical background?

_____

_____

_____

POSSIBLE MATERIALS

research resources about earthquakes
overhead projector
poster board
drawing supplies
printer paper

PLAN THE EXPERIMENT

In your group, decide on the kind of presentation your team will make. Do you want it to be a newscast, part of a magazine show, a documentary, or perhaps a studio forum? What's the purpose of the report? Is the main emphasis geologic or historical? Decide what role each team member will play in the presentation. Table 1 might give you some ideas, but don't limit yourself to the roles there. Each role will require you to research different information. Resources might include the Internet, a library, textbooks, magazine articles, and interviews. How will you decide which of these sources is reliable?

LAB **19.2**  **DESIGN YOUR OWN**

PLAN THE EXPERIMENT, *continued*

Outline your plans and role assignments. Have your teacher check your plans before you start your research.

Use your research notes to draft your section of a script for the presentation and to prepare any visual aids. As a team, read and discuss one another's work. Compile the final script, get your teacher's approval, and rehearse your presentation.

DATA AND OBSERVATIONS

Table 1

Anchor/host	Coordinates the presentation, including introducing and summarizing, introducing experts, and providing emphasis and continuity.
Geologist	Explains technical definitions, such as *tectonics*, *fault*, and *subduction*. Explains what causes earthquakes.
Seismologist	Describes how earthquakes are measured, including instruments, techniques, and magnitude and intensity scales.
Historian	Knows about major historical earthquakes, the destruction that they caused, and their social consequences. Could provide a history of seismology.
Geographer	Concentrates on the location of earthquakes. Explains which regions of the world are prone to earthquakes and provides details of recent earthquakes associated with tsunami devastation.
Disaster worker	Introduces appropriate government agencies and furnishes viewers with practical information about preparation and recovery.

ANALYZE

Take notes as you watch the other teams' presentations. Focus on the three areas described below.

1. Think about the content of the presentation. How comprehensive is the coverage? Is the introduction effective? Does the summary recap the major points well? Are there any major omissions? How could the presenters have improved their coverage?

2. What techniques did the team use to emphasize points? Were minor points overemphasized, or were major points underemphasized?

3. Was the technical content presented in an easily understandable way? Could some points have used more explanation? Were visual aids used effectively?

LAB 19.2   **DESIGN YOUR OWN**

ANALYZE, continued

NOTES

_____

_____

_____

_____

_____

_____

_____

_____

_____

_____

_____

_____

_____

_____

_____

_____

_____

_____

_____

_____

_____

_____

_____

_____

_____

_____

_____

_____

_____

_____

_____

_____

_____

LAB ⟨ **19.2** ⟩   **DESIGN YOUR OWN**

CHECK YOUR HYPOTHESIS

Was your **hypothesis** supported by your data? Why or why not?

_____

_____

CONCLUDE AND APPLY

1. Make a transcript, with illustrations, of your team's presentation.

2. On separate sheets of paper, write a one-paragraph review of each of the other teams' presentations.

3. Would your hypothesis be different now than it was before you developed and presented your script? Explain your answer.

_____

_____

_____

_____

LAB 20.1

INVESTIGATION

Use with
Section 20.1

Plate Tectonics of North America

Changes in the positions and shapes of Earth's continents and oceans can be explained by the theory of plate tectonics. This theory states that Earth's crust and rigid upper mantle are divided into roughly a dozen slabs, called plates. Tectonic plates move slowly over Earth's surface. Interactions among tectonic plates account for most earthquakes, volcanoes, and mountain ranges.

PREPARATION

PROBLEM

How can the theory of plate tectonics be used to analyze some of the tectonic features of North America?

OBJECTIVES

- **Identify** the major plates associated with North America and their movements.
- **Describe** the locations and orientations of major mountain chains of North America.

- **Explain** how geologic evidence supports the theory of plate tectonics.
- **Predict** how future tectonic processes might affect the North American continent.

MATERIALS

red, blue, and orange markers

PROCEDURE

1. Use a blue marker to draw a line on map B that traces the deep-sea trenches off of the western coasts of the Americas.

2. Using a red marker and map A as a reference, draw lines on map B to mark the edges of the tectonic plates shown. Indicate with arrows their directions of movements.

3. Locate the Mid-Atlantic Ridge on map B and color it orange.

4. Study maps B and C. Answer questions 1–5 in Analyze.

5. An active tectonic plate has a leading edge and a trailing edge. On map B, label the leading and trailing edges of the North American Plate.

6. Look at the map legend for map D. What do the dashed lines in the gulf between the Baja Peninsula and the mainland indicate?

   _____

   _____

7. Plate boundaries are convergent, divergent, or transform. Identify each boundary that is associated with the North American Plate. Label them on map B. Answer questions 6 and 7 in Analyze.

   _____

   _____

   _____

LAB ◄ **20.1** ━━━━━━━━━━━━━━━━━━━━━━━━━━ **INVESTIGATION**

DATA AND OBSERVATIONS

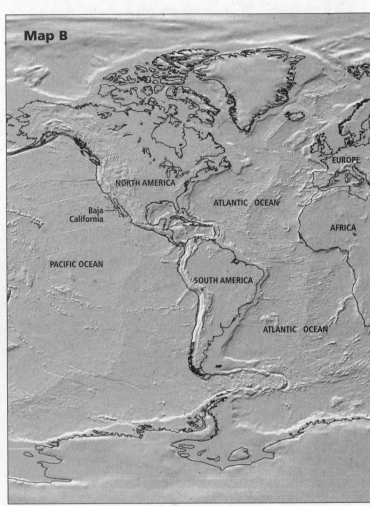

Map B

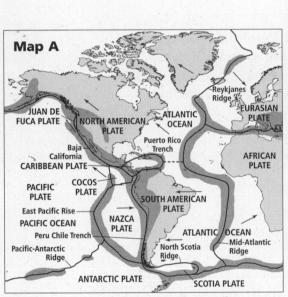

Map A

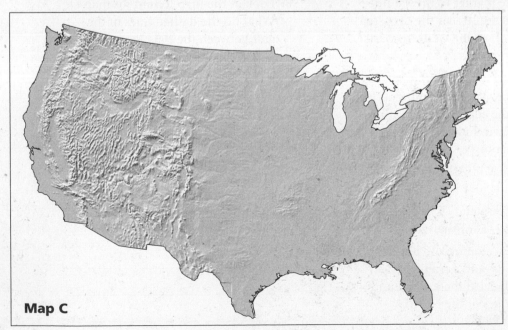

Map C

LAB **20.1**   **INVESTIGATION**

DATA AND OBSERVATIONS, *continued*

Map D

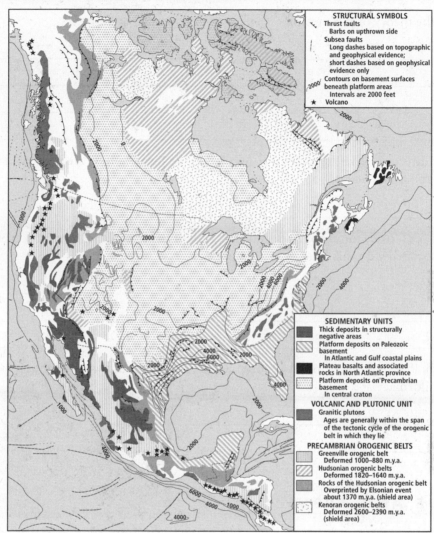

ANALYZE

1. List two features shown on map B that formed or are forming as a result of plate motion.

_____

_____

2. In what direction is the North American Plate moving?

_____

3. Use map C to describe the locations and orientation of the major mountain systems of North America.

_____

_____

_____

LAB **20.1**   **INVESTIGATION**

ANALYZE, *continued*

4. How does the theory of plate tectonics explain your answer to question 3?

_____

_____

_____

_____

5. From what direction were the forces that resulted in the formation of the Appalachian Mountains in the eastern United States?

_____

6. Locate Baja, California, on maps A, B, and D. List all of the tectonic features and events that are associated with this area.

_____

_____

_____

7. If the Pacific Plate continues to move along the San Andreas Fault, what might happen to Baja, California?

_____

_____

_____

CONCLUDE AND APPLY

1. With your group, use the information in maps A–D and what you know about the theory of plate tectonics to briefly describe the tectonic processes that have affected North America.

_____

_____

_____

_____

_____

_____

_____

LAB **20.2**   **MAPPING**

Use with
Section 20.2

Analysis of Geologic Maps

Throughout Earth's history, continents have undergone many structural changes. Most of these changes are due to tectonism, which includes convergence, divergence, folding, faulting, volcanism, and orogeny. Sedimentation and associated subsidence also played a role in changing Earth's continents. Features present at Earth's surface are evidence of the many processes at work to change our planet.

PREPARATION

PROBLEM

How can geologic maps be used to interpret the processes that have resulted in the major landforms of North America?

OBJECTIVES

- **Identify** structural elements of the North American continent by rock age and type.
- **Describe** the tectonic forces that have shaped the mountain ranges of North America.

- **Describe** some of the geologic characteristics of the Appalachian and Rocky Mountain systems.
- **Compare** the tectonic history of some of the major mountain chains of North America.

MATERIALS

map D from Lab 20.1
markers, 8 different colors

PROCEDURE

1. Map D on page 155 is a tectonic map of North America. Carefully study the map and its legend.

2. Find the thick deposits in structurally negative areas and color them yellow.

3. Platform areas are rocks composed of sedimentary or volcanic deposits intruding or overlying older basement rocks. Color the plateau basalts red, the platform deposits on a Paleozoic basement in the Atlantic and Gulf Coastal Plains gray, and the platform deposits on the Precambrian basement in the central craton orange. The craton is the stable interior of the continent. It is composed of large areas of ancient crystalline rocks.

4. Orogenic belts formed as a result of the compression of Precambrian rocks have been identified by the time at which they were deformed. Color the rocks that formed over 2000 million years before present pink. Color the rocks that formed 1820–1640 million years before present green. Color the rocks that formed 1370 million years before present blue. Color the rocks that formed 1000–880 million years before present brown.

5. Color the granitic plutons purple.

6. Find the Valley and Ridge Province of the Appalachian Mountains on Map E. Map F is a close-up of these rocks. On Map F, color the ridges orange and the valleys green.

LAB ◁ **20.2**   **MAPPING**

DATA AND OBSERVATIONS

Map E

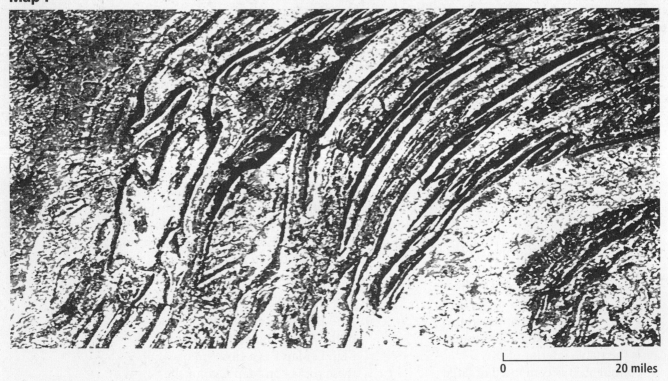

Map F

0 ⊢————————⊣ 20 miles

LAB 20.2   **MAPPING**

ANALYZE

1. What are the yellow areas on map D?

_____

_____

_____

2. On map D, locate and label the shield area. The structural patterns of shield rocks show the roots of a series of deformed mountain belts. Use arrows to indicate on map D the direction of the compressive forces that produced the structural trends in each major portion of the shield.

3. On map D, label the Appalachian and the Rocky Mountain ranges. What is the major difference between these two areas and the central craton and Atlantic and Gulf Coastal Plains?

_____

_____

_____

_____

_____

4. Locate the Valley and Ridge Province of the Appalachian Mountains on maps D and E. Map F shows some of the sedimentary rocks that have been deformed into a series of tight folds and then eroded to form the ridges and valleys. Use arrows to indicate the direction of the stresses that produced this deformation.

LAB  **20.2**   **MAPPING**

CONCLUDE AND APPLY

1. Compare the tectonic activity associated with the formation of the Appalachian Mountains with that of the Rocky Mountains.

_____

_____

_____

_____

_____

_____

_____

2. Which are generally older—the Appalachian Mountains or the Rocky Mountains? Which have been more recently active? Cite evidence from the map for your answer.

_____

_____

_____

_____

_____

_____

_____

LAB **21.1** **INVESTIGATION**

Use with
Section 21.3

Fossilization and Earth's History

Fossils, which are the remains of once-living plants or animals, are important evidence of the physical and environmental history of Earth. Organic remains are preserved by original preservation, permineralization, molds, and casts. Fossils are also sometimes preserved in amber, which formed from the hardened sap of evergreen trees, and by freezing. Some organisms and organism parts have better chances than others of being preserved. Fossil remains that are associated with different geologic time periods provide evidence of how life-forms change over time, as well as information about the location of energy sources such as coal and petroleum.

PREPARATION

PROBLEM

How is evidence of life-forms preserved, and what information is in the fossil record?

OBJECTIVES

- **Construct** models of fossils formed by molding, casting, and original preservation.
- **Compare** the characteristics of different types of fossils.
- **Construct** possible scenarios for fossil formation.
- **Evaluate** the quality of information that comes from the fossil record.

MATERIALS

seashell	freezer
petroleum jelly	dead, unsquashed,
plaster of Paris	hard-bodied insect
water	(such as a beetle
2 plastic spoons	or an ant)
food coloring	waxed paper, about
4 paper cups	10-cm square
2 grapes	liquid glue
marking pen	

SAFETY PRECAUTIONS

- Wear safety goggles and an apron during the lab procedure.
- Wear disposable plastic gloves to protect your hands against the chemicals in the glue.

PROCEDURE

1. Label a paper cup with your names. Fill the cup about two-thirds full of water. Add dry plaster of Paris to the water a little at a time, stirring with a spoon. Stop adding plaster when the mixture is about the consistency of honey.

2. Coat a seashell with a thin layer of petroleum jelly. Press the shell into the surface of the plaster. Allow the plaster to dry overnight.

3. Label two paper cups with your names. Put one grape into each cup. Place one cup in the freezer and one in a warm place where it will not be disturbed. You will examine this experiment in about 1 week.

 **LAB 21.1**   **INVESTIGATION**

PROCEDURE, *continued*

4. In nature, after sediments cover the shell and are cemented and compacted into rock, the shell will often dissolve. Remove the shell from the plaster of Paris to represent this process.

5. Answer Analyze question 1 and draw and label a diagram of what you observe in the plaster of Paris after the shell is removed.

6. Coat the surface of the plaster of Paris with a thin layer of petroleum jelly.

7. Label a paper cup with your names. Fill it about halfway with water. Mix another batch of plaster of Paris, as you did in step 1. Stir a few drops of food coloring into the fresh plaster. Slowly pour the colored plaster into the cup containing the hardened, white plaster. Allow the plaster to dry overnight.

8. After the plaster has dried, carefully tear away the paper cup. Gently pry apart the colored and uncolored blocks of plaster.

9. Draw and label a diagram of what you observe about the colored block of plaster. Answer questions 2–4.

10. Get a dead, unsquashed, hard-bodied insect, a piece of waxed paper, and liquid glue. Put a drop of liquid glue on the waxed paper, then put the insect on the spot of glue. Cover the insect with more glue, and allow the glue to dry solid. Answer question 5.

11. After the grapes have sat for 1 week, examine them. Answer question 6 and the Conclude and Apply questions.

DATA AND OBSERVATIONS

DIAGRAM

ANALYZE

1. What does the dry plaster of Paris look like?

_____

2. What does the colored plaster represent?

_____

3. Which diagram shows a mold? Which shows a cast?

_____

LAB **21.1** **INVESTIGATION**

ANALYZE, *continued*

4. If the shell had been completely buried in sediments and dissolved after the sediment had turned to rock, what kind of impression would it leave in the rock?

_____

_____

5. The first drop of glue on the waxed paper represents sap oozing from a damaged spot in the bark of a tree. In nature, what things have to happen next to result in a fossilized insect encased in amber?

_____

6. How do the two grapes differ in appearance after 1 week?

_____

_____

_____

CONCLUDE AND APPLY

1. Compare the types of information about fossil organisms that can be provided by casts and molds with that of original preservation.

_____

_____

_____

_____

2. What conditions are necessary for the successful preservation of marine organisms?

_____

_____

3. How do low temperatures affect preservation? Why are mammoth fossils that are found in frozen mud often well-preserved?

_____

_____

_____

_____

LAB 21.1   **INVESTIGATION**

CONCLUDE AND APPLY, _continued_

4. How much information about vertebrates (such as ourselves) would be preserved in the fossilization process? What kinds of things that we consider important about our bodies would not usually be preserved? Draw a picture of how a researcher would see you if only half of your bones and teeth and none of your soft parts were preserved and found.

_____

_____

5. Many plants and animals never have a chance to be preserved as fossils. For fossils to provide an accurate representation of the living community, where would marine organisms, insects, and plant materials have to be when they died?

_____

_____

_____

_____

6. Think about plants, animals, and humans living today. If future scientists had to depend upon fossils from our age to interpret our physical and environmental history, what conclusions do you think they would come to? Hint: Which organisms would be preserved? How? Would this be an accurate representation of the living community? Why or why not?

_____

_____

_____

_____

_____

_____

LAB **21.2**   **DESIGN YOUR OWN**

Use with
Section 21.4

Analysis of a Climate-Change Time Line Using Planktonic Foraminifera

Microfossils such as planktonic foraminifera are evidence of how life and environmental conditions have changed during the planet's history. One species of foraminifera, Neogloboquadrina pachyderma, *is an excellent recorder of climatic temperatures through geologic time. When Earth experiences periods of relatively cold temperatures, ocean waters are cooler, and* N. pachyderma *forms a shell that coils to the left. During periods of relatively warm temperatures, when ocean waters are warmer,* N. pachyderma *forms a shell that coils to the right. Populations of these microfossils can be examined and their characteristics used as indicators of climatic change.*

Neogloboquadrina pachyderma (right-coiling)

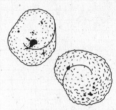

Neogloboquadrina pachyderma (left-coiling)

PREPARATION

PROBLEM

How can a series of foraminifera samples record climatic temperatures?

OBJECTIVES

- **Identify** right-coiling and left-coiling *Neogloboquadrina pachyderma.*
- **Plan** and **carry out** a study of climatic change using *N. pachyderma* as an indicator of temperature change.
- **Gather** microfossil data and **plot** them on a graph.
- **Interpret** the climatic history of Earth during the last 160 000 years.

HYPOTHESIS

As a group, form a hypothesis about how you could use samples containing planktonic foraminifera populations to put together a record of climatic temperatures over geologic time. Write your hypothesis below.

_____

_____

_____

POSSIBLE MATERIALS

representations of microfossil samples, from the present to 160 000 years ago

calculator

PLAN THE EXPERIMENT

Think about how to determine historical changes in climate using the indicator microfossils. Decide how you will analyze the samples. What data will you collect and how will you organize it? What calculations will you have to carry out?

List the steps that you will take for this study. Prepare a table to record the data from each sample, including the percentage of right-coiling *N. pachyderma* present. Select an appropriate style for graphing your

LAB ◆ 21.2   DESIGN YOUR OWN

PLAN THE EXPERIMENT, *continued*

results. Which variables will the graph illustrate? What scale of measurement will you use? How will

the axes be labeled? Have your teacher approve your plan before you begin the study.

_____

_____

_____

_____

_____

_____

DATA AND OBSERVATIONS

LAB **21.2**   **DESIGN YOUR OWN**

ANALYZE

1. What evidence does the data provide about changes in ocean temperature over the years?

_____

_____

_____

2. Do the data contain exact temperatures? Explain your answer.

_____

_____

_____

3. Can you identify any patterns or cycles in the data?

_____

_____

_____

4. What percentage and time intervals did you use for the axes of your graph?

_____

_____

5. How does your graph help you interpret the climatic history of Earth?

_____

_____

CHECK YOUR HYPOTHESIS

1. Was your **hypothesis** supported by your data? Why or why not?

_____

_____

CONCLUDE AND APPLY

1. Describe your conclusion about the climatic history of Earth during the last 160 000 years.

_____

_____

_____

LAB 21.2   **DESIGN YOUR OWN**

CONCLUDE AND APPLY, *continued*

2. What evidence do you have for this conclusion?

_____

_____

_____

3. Is there any evidence from the microfossil data about the causes of the temperature change in the climate time line?

_____

4. Where does Earth appear to be in the climate cycle at present?

_____

5. Review news periodicals, radio and television broadcasts, and nature and science magazines, and talk to your parents and grandparents or other older people you know to find forms of evidence of climate change existing today. Describe two types of evidence and their probable causes.

_____

_____

_____

_____

6. Using the information that you have gathered, predict the direction that the climate-change graph might take from the present into the near future. Explain your prediction.

_____

_____

_____

_____

LAB **22.1**   **INVESTIGATION**

Sequencing Time

Earth has been in existence for 4.6 billion years—a period of time so long it is difficult to imagine. When we discuss a topic such as the reign of the dinosaurs, it may seem like it occurred in Earth's distant past. It may surprise you to know, however, that dinosaurs walked Earth in relatively recent times, geologically speaking. Earth's long history can be represented visually by the geologic time scale, which divides Earth's period of existence into relatively small units. To better appreciate the history of Earth and the passage of geologic time, you will make a graphic version of the geologic time scale and compare it to a time scale of your own life.

PREPARATION

PROBLEM
How do milestones in the existence of Earth correspond to a human lifespan?

OBJECTIVES
- **Compare** the proportionate length of a human life to that of Earth.
- **Calculate** the scale at which Earth's history can be graphically compared to a human lifetime.

- **Analyze** the timing and scope of the evolution and diversification of life on Earth.

MATERIALS
pencil
2.5-m long roll of paper
calculator
meterstick

PROCEDURE

1. On the left side of the roll of paper, construct the geologic time scale. Make 1 cm on your paper equal 20 million years. Include the eons, eras, and geologic time periods shown in the table in Data and Observations. Begin at the bottom of the paper, with the birth of Earth 4600 million years ago.

2. Now construct a personal time scale. To do this, you will map your life alongside the length of existence of Earth. The scale will be different from the scale you used for the geologic time scale in step 1. Begin by computing the proportionate length of one year of your life.

The formula for this step is shown below.

> Mapped length of one year of life = total geologic time scale (in cm)/number of years in your life (to the closest month; expressed as a decimal)

Recall that mapping the 4600 million years of Earth's existence took 230 cm, so for a person who is 15 years and 8 months old: Mapped length of one year of life = 230 cm/15.75 years = 14.6 cm/year of life

LAB 22.1   **INVESTIGATION**

PROCEDURE, *continued*

3. On the right side of the roll of paper, mark off the years of your life using the scale you computed. Your birth should begin alongside the birth of Earth. The current time in your life should fall alongside the current time in Earth's history noted at the top of the roll of paper.

4. Refer again to the table and in the center of the paper, between the two scales you have drawn, write in important events in the evolution of life on Earth. For example, in the section that corresponds to the Ordovician, write "Fishes," and so on.

DATA AND OBSERVATIONS

Table

Era or Eon	Period	End Date (millions of years ago)	Length (in millions of years)	Life on Earth
Cenozoic	Quaternary	—	2	Humans
	Neogene	1.6	64	Mammals diversify.
	Paleogene	23		
Mesozoic	Cretaceous	66	80	Dinosaurs become extinct.
	Jurassic	146	62	Birds
	Triassic	208	37	Dinosaurs, mammals
Paleozoic	Permian	245	45	Seed plants
	Pennsylvanian (Carboniferous)	290	33	Reptiles diversify.
	Mississippian (Carboniferous)	323	39	Reptiles
	Devonian	362	46	Amphibians
	Silurian	408	31	Land invertebrates and plants
	Ordovician	439	71	Fishes
	Cambrian	510	30	Marine invertebrates
Proterozoic Eon (Precambrian)		540	1960	Multicellular organisms
Archean (Precambrian)		2500	2100	Early bacteria
		4600		Birth of Earth

LAB **22.1**   **INVESTIGATION**

ANALYZE

1. What was the longest division of time in Earth's history? Approximately what percentage of Earth's history occurred during this division?

   _____

   _____

   _____

   _____

2. Relate this long division of Earth's existence to your own life. About how many years of your life is proportional to the longest division of time in Earth's history?

   _____

   _____

   _____

3. About what percentage of the entire existence of Earth have bacteria been alive?

   _____

   _____

   _____

   _____

   _____

LAB ⟨**22.1**⟩   **INVESTIGATION**

CONCLUDE AND APPLY

1. What can you conclude about the rise of living things? Did the rise of new organisms and the diversification of living things occur at an even rate over the entire history of Earth?

   _____

   _____

   _____

   _____

   _____

2. Assuming that humans have been alive for 2 million years, to what length of time in your own life is this proportional?

   _____

   _____

   _____

   _____

   _____

   _____

LAB **22.2**   **MAPPING**

Use with
Section 22.4

What came first?

Evidence of early life-forms and the planetary conditions that existed during the Precambrian provides us with information that we can use to make inferences about changes in the crust, atmosphere, and oceans throughout Earth's history. These changes contributed to the conditions necessary for the development of more advanced life-forms and the plant and animal life that exists today.

PREPARATION

PROBLEM

How can we illustrate the relationship of Precambrian life-forms to those that exist today?

MATERIALS

geologic time scale
event card

OBJECTIVES

- **Represent** life-forms that appeared during different periods of Earth's history, beginning with the Precambrian.
- **Relate** time to the number and complexity of organisms on Earth.
- **Describe** how information about the Precambrian can be used to **analyze** planetary materials in the search for life elsewhere in the universe.

PROCEDURE

1. Examine the time scale that has been hung the room. Discuss its characteristics with your group members and the class.

2. Pick an event card. Each card represents an event or the first evidence of an organism in the fossil record. You will become that organism or event. (Refer to Table 1 for your approximate geologic time.) As your approximate age is announced, stand in front of the time scale at the place that represents that age. If your organism becomes extinct, sit on the floor when the Permian Mass Extinction Event or the Cretaceous-Paleogene Mass Extinction Event is announced.

LAB 22.2 MAPPING

DATA AND OBSERVATIONS

Event or First Evidence in Fossil Record	Approximate Geologic Time
Bacteria	3 b.y.b.p.
Green algae	1 b.y.b.p.
Jellyfish	600 m.y.b.p.
Ediacara organisms	550 m.y.b.p.
Eurypterids*	510 m.y.b.p.
Horn corals*	500 m.y.b.p.
First vertebrates	480 m.y.b.p.
Spiders	400 m.y.b.p.
Sharks	400 m.y.b.p.
First jawed fish*	380 m.y.b.p.
Ferns	350 m.y.b.p.
Earthworms	300 m.y.b.p.
Great Permian Extinction Event	250 m.y.b.p.
Drastic geographic and climatic changes	248 m.y.b.p.
First dinosaurs*	220 m.y.b.p.
First mammals	210 m.y.b.p.
Ginkgo biloba	200 m.y.b.p.
Abundant ammonites*	180 m.y.b.p.
*Archaeopteryx**	140 m.y.b.p.
First flowering plants	120 m.y.b.p.
Ants	100 m.y.b.p.
Cretaceous-Paleogene Extinction Event	66 m.y.b.p.
Camel	35 m.y.b.p.
Grass	20 m.y.b.p.
Australopithecus afarensis ("Lucy")	4 m.y.b.p.

* = now extinct
b.y.b.p. = billion years before present
m.y.b.p. = million years before present

LAB **22.2**   **MAPPING**

ANALYZE

1. List the major divisions of Earth's history that are represented on this time scale.

_____

2. Identify the approximate beginning and ending points of these major divisions on the time scale, in inches.

_____

_____

_____

3. Which of these divisions is the longest? Shortest? Oldest? Most recent?

_____

_____

_____

4. From your observations of the completed time scale, describe two major differences between the Precambrian and the Cenozoic Era. Why do you think these differences exist?

_____

_____

_____

_____

_____

_____

LAB **22.2**   **MAPPING**

CONCLUDE AND APPLY

1. Evidence from the Precambrian indicates that organisms at that time produced the oxygen that changed the composition of the atmosphere. What effect did this have on the evolution of life-forms?

_____

_____

_____

_____

2. What were the most important overall impressions that you got from observing the construction of the time scale?

_____

_____

_____

_____

3. How might information from the Precambrian be useful in analyzing extraterrestrial material for evidence of life elsewhere in the universe?

_____

_____

_____

_____

LAB ◆ **23.1**   **MAPPING**

Water to Land

**Use with
Section 23.2**

Early life-forms on Earth developed in the seas. As the climate and the shapes, sizes, and locations of these seas and landforms changed, some life-forms developed adaptations that enabled them to move onto land. Paleontologists are able to reconstruct Earth as it was millions of years ago by studying fossils to infer what environments were present at different times.

PREPARATION

PROBLEM

How can we use representative fossils to infer the type of environment in which organisms once lived?

MATERIALS

fossil cards or models
blue, yellow, and green pencils

OBJECTIVES

- **Map** the fossils in a progression of different environments.
- **Categorize** fossils based on adaptations for survival in sea, beach, or land environments.
- **Draw** the boundaries of environments on a fossil map.
- **Defend** interpretations of fossil evidence.

PROCEDURE

1. Examine a fossil card or model at a station.

2. In the space in Data and Observations that has the same number as the fossil, sketch the fossil and label it. You will have about 2 minutes for this.

3. Move to the next station and repeat steps 1 and 2, until you have sketched all the fossils.

LAB **23.1**   **MAPPING**

DATA AND OBSERVATIONS

1	2	3	4	5
6	7	8	9	10
11	12	13	14	15
16	17	18	19	20
21	22	23	24	25
26	27	28	29	30

LAB ◆ **23.1** MAPPING

ANALYZE

1. What does this fossil evidence indicate about the types of environments in which the organisms might have lived?

_____

_____

_____

2. Examine the distribution of the fossils in Data and Observations. Using dashed lines, indicate where one environment might have ended and another environment might have begun. Describe how the environments are arranged.

_____

_____

_____

_____

_____

LAB ◄ **23.1** ►   **MAPPING**

CONCLUDE AND APPLY

1. Color your filled-in grids as follows:

boxes with land fossils = green

boxes with beach fossils = yellow

boxes with sea fossils = blue

What are the major differences among the organisms that lived in these different environments?

_____

_____

_____

_____

_____

2. Because you did not actually observe these organisms while they were living in the three environments, you have inferred their habitats from fossil evidence. List the reasons for your inferences of the evidence.

_____

_____

_____

_____

_____

_____

_____

_____

_____

LAB 23.2

MAPPING

Cenozoic Ice Sheets and Plant Distribution

An analysis of pollen grains from the Cenozoic Era can provide evidence of past climatic conditions. Because we know the conditions in which many plant species grow today, changes in ancient climates can be estimated by comparing fossil pollen samples with living relatives and estimating ancient plant distribution from the changing concentrations of pollen remains. Changes in climate can then be inferred from the changes in the distribution of vegetation.

PREPARATION

PROBLEM

How can the relationship between changes in the extent of the Laurentide ice sheet in North America and changes in the distribution of vegetation inform us about climate change during the Cenozoic?

OBJECTIVES

- **Describe** the distribution of different plant groups in North America at different times during the last ice age.
- **Explain** the relationship between changes in plant distribution and the extent of the Laurentian ice sheet.
- **Make inferences** about the relationship between plant distribution and climate change.

MATERIALS

fine-point colored markers or
   colored pencils
political map of North America

PROCEDURE

The maps on the next page show the distribution of pollen in the Eastern United States from the time of the last ice age, the Laurentide, to the present. The label ka means thousands of years ago. Refer to the legend and use the markers to color the North American plant distribution maps as follows:

Plant Group	Color
Ice	White
Tundra	Violet
Forest Tundra	Purple
Boreal Forest (northern forest south of the tundra)	Dark Green
Mixed Forest	Light Green
Deciduous Forest	Orange
Aspen Parkland	Yellow
Prairie	Brown
Southeast Forest	Blue
No Match	Red
No Data	Gray

LAB 23.2   **MAPPING**

DATA AND OBSERVATIONS

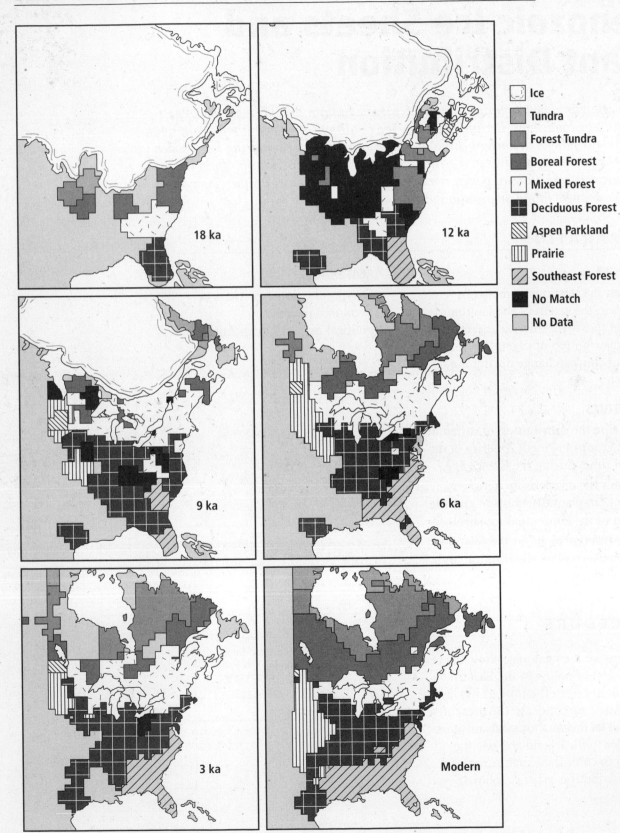

LAB **23.2**   **MAPPING**

ANALYZE

1. The extent of the Laurentide ice sheet has varied over the past 18 000 years.

    a. At what time was the Laurentide ice sheet at its greatest extent?

    _____

    b. How many plant groups were present at this time? What were they?

    _____

    _____

    _____

    c. By what approximate time was the ice sheet first absent from the North American continent?

    _____

    d. How many plant groups were present at this time? What were they?

    _____

    _____

    _____

    e. What information does this change in number and type of plant groups give us about climate?

    _____

    _____

    _____

2. At 18 ka, tundra vegetation was present in the area that today is the Midwest.

    a. By 9 ka, what type of vegetation had replaced the tundra?

    _____

    b. What can we infer about climate from this change in vegetation?

    _____

    _____

3. Tundra now exists in northern Canada, Alaska, and Siberia.

    a. What do we know about today's climate in those areas?

    _____

    b. Which states today coincide with the approximate area that was covered by tundra at 18 ka?

    _____

LAB ◀ **23.2** ▶   **MAPPING**

ANALYZE, continued

c. According to the maps, Florida's climate at 18 ka was similar to today's climate in which of today's areas?

_____

_____

CONCLUDE AND APPLY

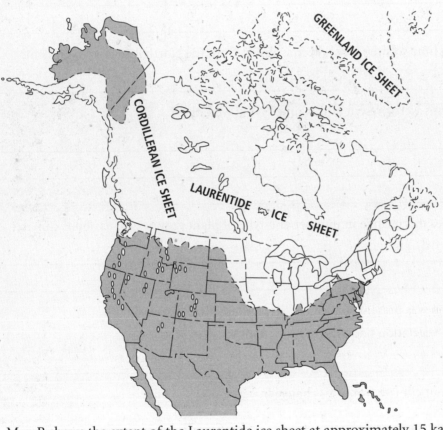

1. Map B shows the extent of the Laurentide ice sheet at approximately 15 ka. Using what you know about the relationship of climate to plant group, predict the plant group distribution for the western half of North America at that time. Color those areas on the map. Explain your prediction.

_____

_____

_____

_____

_____

_____

LAB ◆ **24.1** **DESIGN YOUR OWN**

Neutralizing Acid Precipitation

W hen fossil fuels are burned to power vehicles or to generate electricity, they produce sulfur dioxide and nitrogen oxides in the atmosphere. These pollutants combine with atmospheric moisture to create sulfuric acid and nitric acid. As a result, the precipitation is more acidic than unpolluted forms of precipitation. Acid precipitation can decrease soil productivity and damage buildings, vegetation, and wildlife.

PREPARATION

PROBLEM

How can the acidity of acid precipitation be decreased, or neutralized?

OBJECTIVES

- **Design** an experiment to **test** the pH levels of two solutions.
- **Determine** how to neutralize an acidic solution.
- **Compare** and **contrast** an acidic solution and a neutralized solution.

HYPOTHESIS

As a group, review the pH scale and the characteristics of acidic and basic solutions. Write a hypothesis about how an acidic solution could be neutralized. Write your hypothesis below.

_____

_____

_____

POSSIBLE MATERIALS

500 mL distilled water
230 g powdered limestone
white vinegar
baking soda
250-mL glass beakers (2)
stirring rods
measuring spoons
dropper
marking pen
plastic wrap
pH paper
pH color chart

SAFETY PRECAUTIONS

- Wear safety goggles, gloves, and an apron during the lab procedure.
- If you break any glassware, notify your teacher right away. Do not attempt to clean up broken glass.
- Label all solutions.
- Follow your teacher's suggestions for disposing of lab materials.

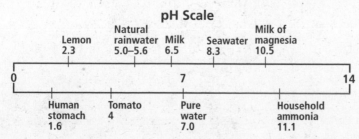

pH Scale

Lemon 2.3 · Natural rainwater 5.0–5.6 · Milk 6.5 · Seawater 8.3 · Milk of magnesia 10.5

Human stomach 1.6 · Tomato 4 · Pure water 7.0 · Household ammonia 11.1

LAB 24.1   **DESIGN YOUR OWN**

PLAN THE EXPERIMENT

Review the list of possible materials. Design a plan
to slowly decrease, or neutralize, the acidity of a
solution over the course of several days. Your
experiment should include a control, a way to test
pH, and a method to ensure that the solutions do not
evaporate. Set up a table to record your results. The
pH color chart provided by your teacher and the pH
scale will help you interpret your results. Outline
your plan and have your teacher approve it before
you begin the experiment.

DATA AND OBSERVATIONS

DATA TABLE

LAB **24.1** **DESIGN YOUR OWN**

DATA AND OBSERVATIONS, *continued*

GRAPH

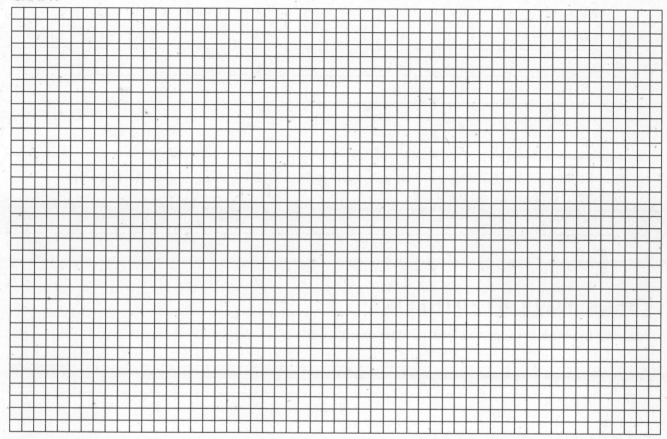

ANALYZE

1. What was your control? What was the independent variable in this experiment?

_____

_____

2. How did you decrease the acidity of your solution? Why was your method effective?

_____

_____

_____

_____

LAB 24.1   DESIGN YOUR OWN

ANALYZE, *continued*

3. Plot the pH levels of the solutions on a graph below your table. Compare and contrast the pH levels of the two solutions. Describe how these levels changed during the experiment.

_____

_____

_____

_____

_____

4. Plot the final pH levels of the solutions on the pH scale in Figure 1. How do these levels compare to those of other items on the scale?

_____

_____

CHECK YOUR HYPOTHESIS

Was your **hypothesis** supported by data? Why or why not?

_____

_____

CONCLUDE AND APPLY

1. Suppose that the pH levels of your solutions represent the pH levels of two separate rainfalls. How might each rainfall affect vegetation and aquatic organisms?

_____

_____

_____

2. Based on your results, describe how acid precipitation can be neutralized in lakes and ponds.

_____

_____

3. Would the method you described in question 2 solve the problem of acid precipitation permanently or temporarily? Explain your answer.

_____

_____

LAB ◄ **24.2** **INVESTIGATION**

Use with
Section 24.4

Water Usage

According to the United States Geological Survey (USGS), the government agency that oversees United States water supplies, the United States has abundant freshwater resources. However, many existing sources of groundwater and surface water, such as wells, lakes, and reservoirs, are in danger of being overused, and drought is a problem in some areas. The economic and environmental health of the country—and of the entire planet—depends, in part, on maintaining a balance between water demand and water supply.

PREPARATION

PROBLEM
How has water usage changed in the United States since 1950?

OBJECTIVES
- **Analyze** changing trends in water usage over a 40-year period.

- **Determine** which categories use the most water per day.
- **Discuss** conservation methods that might decrease water use.

MATERIALS
calculator

PROCEDURE

1. Study the bar graph, based on data from the USGS. Look for patterns in water usage. Do certain categories consistently use more water than others? Have some categories increased their water use more quickly than others?

2. Create a data table based on the graph. Place the nine time periods across the top of the table. List the

five categories on the left side of the table. In the rows, write the total amount of water used per day by each category for the years listed.

3. Calculate the total amount of water used per day by all categories for each year. Add this information to the data table.

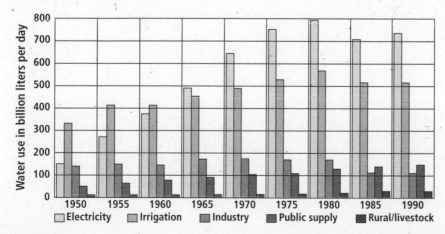

LAB 24.2   **INVESTIGATION**

DATA AND OBSERVATIONS

Data Table

LAB ◁ **24.2** ▷ **INVESTIGATION**

ANALYZE

1. Approximately how many billions of liters of water were used per day in the United States in 1990? How does this amount compare to total water usage in 1950?

_____

_____

2. Which category showed the greatest increase over the course of the 40 years shown in the graph? What might account for the increase?

_____

_____

_____

3. Which category showed a decrease in water use over the course of 40 years? What might account for the decrease?

_____

_____

_____

4. When did water usage peak in the United States for most categories? Which categories continued to show an increase? Make a hypothesis for the reasons behind the increase.

_____

_____

_____

5. Which two categories consistently account for most water usage in the United States? How much water did these categories use in 1980?

_____

_____

6. Describe how water might be used for each category in the graph.

_____

_____

_____

_____

LAB **24.2**    **INVESTIGATION**

CONCLUDE AND APPLY

1. The 1990 population of the United States was about 252 million. Use the amount listed in the "Public supply" category for 1990 to calculate the average amount of water each United States citizen used daily.

_____

_____

_____

2. In 1995, total United States water use was about 1520 billion L per day. Add this information to your data table. In your own words, describe the overall trend in water usage in the United States from 1950 to 1995.

_____

_____

_____

_____

3. Think of all the ways that you use water each day. Describe at least three things that you could do as an individual to decrease your daily water use. Identify two things that government or industry could do to decrease water use.

_____

_____

_____

_____

LAB 25.1 DESIGN YOUR OWN

Use with Section 25.2

Solar Water-Heater

Scientists estimate that, in one hour, enough solar energy reaches Earth to meet global energy needs for an entire year. Solar energy is nonpolluting, renewable, and cost effective. Why isn't it more widely used? Because solar energy is not available at night or on overcast days, it must be stored in some way. Also, many areas do not receive enough sunlight to make solar energy a viable large-scale resource. Technological advances will likely solve these problems in the near future. Meanwhile, there are many small-scale uses for solar energy.

PREPARATION

PROBLEM

What factors must you consider when building a solar water-heater?

OBJECTIVES

- **Design** and **construct** a solar water-heater.
- **Analyze** the efficiency of various solar energy devices and **suggest** improvements to them.
- **Discuss** the advantages and disadvantages of solar energy.

HYPOTHESIS

As a group, discuss how you could use the materials provided by your teacher to design and build a solar water-heater. Form a hypothesis about the factors that might affect the efficiency of your solar energy device.

_____

_____

_____

_____

_____

_____

POSSIBLE MATERIALS

3 L water

3 m black tubing

shallow cardboard box, about
  30 cm × 40 cm

black construction paper

clear plastic wrap

insulating materials
  (cotton, newspaper, cloth, etc.)

tape

scissors

alcohol-based thermometer

clothespin

2 clean plastic 2-L bottles

SAFETY PRECAUTIONS

- Wear safety goggles during the lab procedure.
- If you break a thermometer, notify your teacher right away. Do not attempt to clean up broken glass.
- Use caution when handling sharp objects such as scissors.
- Avoid using mercury-based thermometers. Mercury is toxic.
- Follow your teacher's suggestions for disposing of lab materials.

LAB 25.1   **DESIGN YOUR OWN**

PLAN THE EXPERIMENT

Review the list of possible materials. Working individually, design a device to warm water by using solar energy. Sketch your design below and label its parts. With your group, decide upon the best design. You may wish to incorporate elements from several designs. Sketch and label the revised design below. Plan how to test the design's effectiveness. How will you ensure that the heater receives enough solar energy? What will you use for insulation? How will you maintain a steady flow of water through the heater? Set up a data table to record your results. Have your teacher approve your plan before you build and test the heater.

DATA AND OBSERVATIONS

DATA TABLE

LAB **25.1** DESIGN YOUR OWN

DATA AND OBSERVATIONS, *continued*

GRAPH

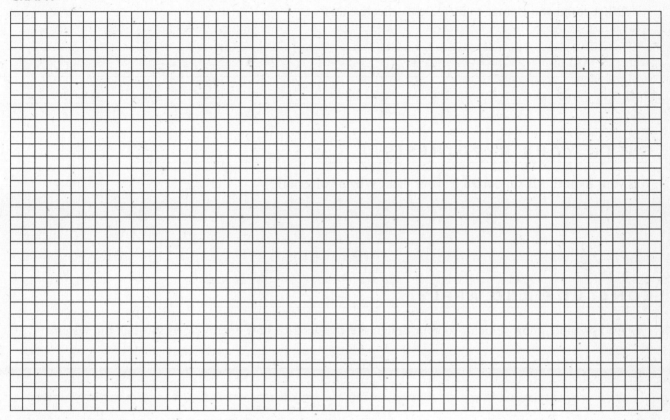

ANALYZE

1. How did you test the effectiveness of your solar water-heater?

_____

_____

_____

2. Graph the data that you gathered as you tested your device.
Describe any patterns that you observe in the data.

_____

_____

_____

3. Did your solar water-heater function as planned? Explain your answer.

_____

_____

LAB **25.1** **DESIGN YOUR OWN**

ANALYZE, *continued*

4. Compare your designs to those of other groups. Think about improvements to your design. What could you change to increase its effectiveness?

_____

_____

_____

CHECK YOUR HYPOTHESIS

Was your **hypothesis** supported by your data? Why or why not?

_____

_____

CONCLUDE AND APPLY

1. Add possible improvements to your design sketch. Describe how energy flows through the system.

_____

_____

_____

2. Review your results and those of other groups. Which factors appear to affect the efficiency of solar water-heaters?

_____

_____

_____

3. Based on what you have learned in this activity and on your previous knowledge of solar energy, identify two advantages and two disadvantages of solar energy.

_____

_____

_____

LAB ◄ **25.2** ►   **INVESTIGATION**

Assessing Wind Energy

Humans *have used energy from the wind for thousands of years to pump water and grind corn. Experts estimate that this nonpolluting source of energy may provide up to 10 percent of electricity needs worldwide within the next 50 years. Wind energy is generally produced at wind farms that use wind turbines to turn generators and create electricity. Currently, however, electricity from wind energy is difficult to store and to transport long distances. Until technology improves, wind energy is largely limited to areas with steady, strong winds.*

PREPARATION

PROBLEM
Is wind energy a viable energy resource for your area?

OBJECTIVES
- **Construct** a tool to measure wind speed.
- **Observe** and **record** wind speeds at different locations.
- **Determine** if local wind speeds are high enough to generate electricity.
- **Consider** the advantages and disadvantages of wind energy.

MATERIALS
small plastic ball
white paper
tape
cardboard, 10 cm ✕ 16 cm
scissors
marking pen
heavy-duty sewing needle
heavy thread, 30 cm
calculator

SAFETY PRECAUTIONS

- Wear safety goggles during the lab procedure.
- Be careful not to stab yourself when punching a hole with the needle.
- Use caution when handling sharp objects such as scissors.

PROCEDURE

1. Use white paper and a pencil to trace Figure 1. Cut out the paper protractor and tape it to the cardboard.

2. Trace the outline of the paper protractor on the cardboard, and cut it out. Mark the angles of the protractor, as shown in Figure 1.

3. Thread the sewing needle with heavy thread. Carefully use the threaded needle to punch a hole through the center of a small plastic ball. The needle should go completely through the ball. Unthread the needle and tie a knot in the thread so the ball will not slide off the end.

PROCEDURE, *continued*

Figure 1

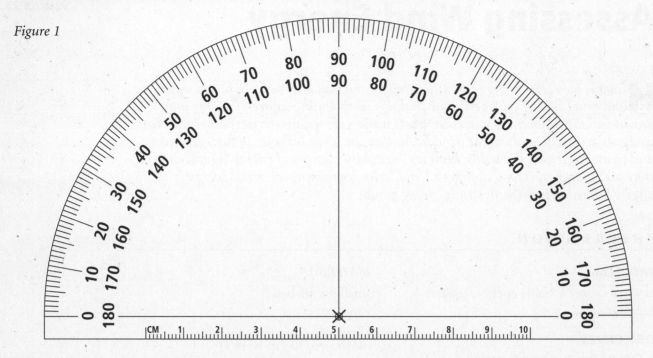

4. Tape the other end of the thread to the cardboard protractor. The *X* on the protractor in Figure 1 indicates where to tape the thread.

5. Go outside and choose three sites near your school that are relatively open. Stand with your back to the wind and hold the protractor level with the straight edge up so that the thread with the ball hangs down parallel to the 90° mark. Hold the protractor in front of you so that you do not completely block the wind. The thread with the ball should be able to move freely along the face of the protractor.

6. Measure the angle of the thread on the protractor (see Figure 2). Record the angle in Table 2. Use Table 1 to convert your measurements into actual wind speeds.

7. Measure and record wind speeds at the other two sites. Collect data at each site daily for 5 days.

Figure 2

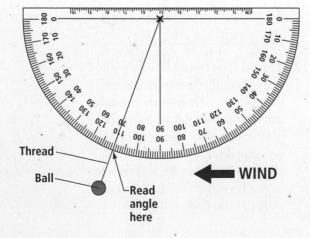

Thread

Ball

Read angle here

← **WIND**

LAB  **25.2**   **INVESTIGATION**

DATA AND OBSERVATIONS

Table 1

Angle (°)	Approximate Wind Speed (km/h)	Angle (°)	Approximate Wind Speed (km/h)	Angle (°)	Approximate Wind Speed (km/h)	Angle (°)	Approximate Wind Speed (km/h)
90	0	70	12	50	18	30	26
85	6	65	13	45	20	25	29
80	8	60	15	40	21	20	33
75	10	55	16	35	23		

Table 2

	Day	Site 1	Site 2	Site 3
Angle	1			
Wind speed	1			
Angle	2			
Wind speed	2			
Angle	3			
Wind speed	3			
Angle	4			
Wind speed	4			
Angle	5			
Wind speed	5			

ANALYZE

1. Calculate the average wind speed at each site. Compare your averages to those of other groups. Did wind speeds vary widely or were they relatively constant for all groups? Give reasons for the similarity or variation.

_____

_____

_____

_____

2. Where were the greatest wind speeds measured? Describe the layout of those areas.

_____

_____

LAB **25.2**   **INVESTIGATION**

ANALYZE, continued

3. Use the space below to plot your average wind speeds. Describe any daily variations in the plotted values. How might these variations affect the reliability of wind as an energy resource?

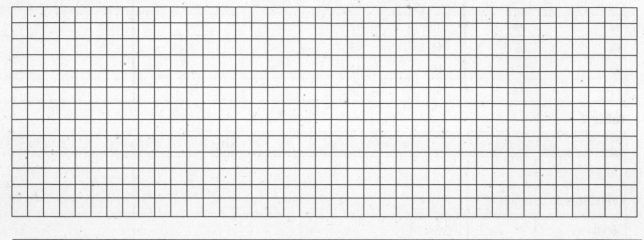

_____

_____

_____

4. To generate electricity from wind efficiently, average wind speeds must be at least 21 km/h. How do your average values compare to this value?

_____

_____

CONCLUDE AND APPLY

1. Wind energy increases with the cube of the wind speed. If local average wind speed was 10 km/h, how much of an increase in wind speed would be necessary to double local energy output?

_____

_____

2. An area that experiences weak winds is not suitable to generate electricity by wind. An area that experiences very strong winds is also not suitable to generate electricity by wind. Why?

_____

_____

3. Based on your measurements of wind speed, would your area be a suitable place for the production of electricity by wind energy? Explain your answer.

_____

_____

LAB ◆ **26.1**   **DESIGN YOUR OWN**

Use with
Section 26.2

Cleaning Up Oil Spills

Extracting fossil fuels such as coal from Earth's crust often results in pollution of land or water resources. Pollution may also occur during transport of fossil fuels. Oil spills caused by tanker collisions or leaks at offshore oil wells may form thick layers of crude oil on the water's surface, which can interfere with marine organisms' ability to feed, breathe, move, and reproduce. Major oil spills like the one from the Exxon Valdez in Prince William Sound have disrupted food webs and devastated marine ecosystems.

PREPARATION

PROBLEM

Which materials are most effective in cleaning up an oil spill?

OBJECTIVES

- **Develop** criteria to **determine** the effectiveness of different materials in cleaning up oil spills.
- **Model** a cleanup of an oil spill.
- **Compare** and **contrast** the effectiveness of different materials as cleaning agents for oil spills.

HYPOTHESIS

As a group, decide how various materials could be used to clean up an oil spill. Form a hypothesis about which material will be the most effective cleaning agent. Write your hypothesis below.

_____

_____

_____

POSSIBLE MATERIALS

olive oil
shallow pan
paper towels
water
toothpicks
feathers
string
cotton balls
cardboard
dropper
liquid detergent
gravel
sponge

SAFETY PRECAUTIONS

- Wear safety goggles, an apron, and gloves during the lab procedure.
- Dispose of oil-soaked materials as directed by your teacher.

LAB 26.1   **DESIGN YOUR OWN**

PLAN THE EXPERIMENT

As a group, discuss and agree upon a plan stating
how you could clean up an oil spill using the
materials provided. Write down your plan. Be
specific about how you will use the gravel and the
pan to model a rocky beach. How will you model the
oil spill? You will need to design and conduct several
tests to determine the most effective cleaning agents.
In your plan, explain how you will separate and
control variables, so that you test only one variable
at a time. Set up a table to record your results. Have
your teacher approve your plan before you begin the
experiments.

_____
_____
_____
_____
_____
_____
_____

DATA AND OBSERVATIONS

DATA TABLE

LAB ◁ **26.1** ▷   **DESIGN YOUR OWN**

ANALYZE

1. Which materials were most effective in absorbing the oil from the water's surface?
Which materials were least effective?

_____

_____

2. What criteria did you use to judge how well the materials cleaned up the oil spill?

_____

_____

3. The water in the pan represented the open ocean, and the gravel represented a rocky shore.
Did the cleaning agents perform equally well in both locations? Why or why not?

_____

_____

_____

_____

4. Identify some disadvantages that may be associated with your cleaning materials. For instance,
are any of your cleaning methods harmful to marine organisms?

_____

_____

_____

5. Based on your results, would you recommend a combination of cleaning methods or one
method alone? Explain your answer.

_____

_____

_____

LAB   **26.1**   **DESIGN YOUR OWN**

CHECK YOUR HYPOTHESIS

Was your **hypothesis** supported by your data? Why or why not?

_____

_____

CONCLUDE AND APPLY

1. Describe the appearance of the feathers after they were dipped in the oil. Based on your observations, predict how an actual oil spill might affect a seabird.

_____

_____

_____

2. Oceans are in constant motion. Describe some of these movements. Would the motion of the water help or hinder efforts to clean up an oil spill?

_____

_____

_____

LAB ◄ **26.2** ▶   **INVESTIGATION**

Algal Blooms

Farmers commonly add fertilizers containing nitrogen, potassium, and phosphorus to their fields to increase crop production. These chemicals can leach into the soil and run off into nearby ponds, streams, and lakes. When an excessive amount of fertilizer enters freshwater, algae and aquatic plants can experience population explosions. If the algal population explodes and covers the surface of the pond, it is called an algal bloom.

PREPARATION

PROBLEM

Under what conditions do algae grow best?

OBJECTIVES

- **Observe** the growth of algae in two controlled experiments.
- **Discover** what makes algae thrive.
- **Recognize** that excessive growth of algae may be linked to human activities.

MATERIALS

2 L distilled water
2 L liquid fertilizer
1 L pond water
1-L glass jars with lids (4)
100-mL graduated cylinder

marking pen
microscope
microscope slides
coverslips
dropper

SAFETY PRECAUTIONS

- Wear splash-resistant safety goggles, gloves, and an apron during the lab procedure.
- If you break any glassware, notify your teacher right away. Do not attempt to clean up broken glass.
- Follow your teacher's suggestions for disposing of lab materials.

PROCEDURE

1. Use a marking pen to label four jars from 1 to 4.

2. Measure 475 mL of distilled water into each jar. Add 250 mL of pond water to each jar.

3. Carefully measure 5 mL of liquid fertilizer into jar 2 and into jar 4. Screw the lids on all four jars.

4. Place jars 1 and 2 (experiment A) on a sunny windowsill. Place jars 3 and 4 (experiment B) in a cool, shady area of the lab where they will not be disturbed.

5. Each day for 10 days, observe any changes in the water in the jars. Record your observations in the table provided.

6. On the last day of the experiments, put a drop of water from jar 1 on a microscope slide and cover it with a coverslip. Look at the slide under a microscope. Record your observations. Observe water from the other three jars as well.

LAB ◆ 26.2   INVESTIGATION

DATA AND OBSERVATIONS

	Observations			
	Experiment A		Experiment B	
Day	Jar 1	Jar 2	Jar 3	Jar 4
1				
2				
3				
4				
5				
6				
7				
8				
9				
10				

ANALYZE

1. What was the independent variable for experiment A? What was the independent variable for experiment B? Identify the control in both experiments. How did experiment A differ from experiment B?

_____

_____

_____

_____

_____

2. Describe any changes that occurred in the water in the jars. Did the color of the water change in all four jars? What did you observe under the microscope?

_____

_____

_____

_____

_____

LAB ◁ **26.2** ▷ INVESTIGATION

ANALYZE, continued

3. Which jar showed the greatest algal growth? Which jar showed the least growth? Explain why.

_____

_____

_____

_____

_____

_____

4. Explain why the pond water was a crucial component of this lab.

_____

_____

_____

_____

5. Based on your observations, under which conditions would algae grow best?

_____

_____

_____

_____

_____

LAB 26.2   **INVESTIGATION**

CONCLUDE AND APPLY

1. Algae use oxygen during respiration and decomposition. That means a large population of algae can deplete the oxygen in a body of water. How might this affect fish and other aquatic life in the water?

_____

_____

_____

_____

2. Fertilizers sometimes run off into ponds and cause algal blooms. Describe how a farmer could try to lessen this problem.

_____

_____

_____

_____

3. Based on what you have learned about algal growth, why do algal blooms sometimes occur near power plants and factories that release hot water into a river or lake?

_____

_____

_____

_____

_____

LAB ◄ **27.1** ►   **INVESTIGATION**

Use with
Section 27.1

Make Your Own Telescope

The earliest telescopes were refracting telescopes, which have two lenses. The large one at the front is the objective, and the small one that you look through is the eyepiece. Lenses have two main properties—size and power. The size of a lens is its diameter. The power of a lens depends on its focal length. When you use a lens to project an image on a screen, the focal length is the distance of the image from the lens.

PREPARATION

PROBLEM
Which combination of lenses will make the best refracting telescope?

OBJECTIVES
- **Measure** the diameter and focal length of lenses.
- **Find** the ideal telescope length, given a pair of lenses whose focal lengths are known.
- **Examine** the magnification properties of various pairs of lenses.
- **Construct** a telescope.

MATERIALS
set of 3 lenses with long focal lengths
   (set A: A1 has the shortest focal length of the set; A3 has the longest; and A2 is in between)
set of 3 lenses with short focal lengths
   (set B: B1 has the shortest focal length of the set; B3 has the longest; and B2 is in between)

set of 3 lenses of identical focal length
   and different diameters (set C)
ring stand
burette clamp
meterstick
2 lens holders
screen
screen holder
2 nested cardboard tubes
   (combined length 1 m)
2 foam lens holders

SAFETY PRECAUTIONS

- Under no circumstances should you look at the Sun through a telescope; it could cause permanent damage. Do not project images of the Sun through a lens. Wear goggles to help decrease any glare.
- Be careful when handling glass lenses. Watch out for sharp edges.

PROCEDURE

1. Measure and record in Table 1 the diameter of each lens to the nearest tenth of a centimeter.

2. Use a burette clamp to attach a meterstick to a ring stand. The meterstick should be horizontal.

This arrangement is known as an optical bench. Mount the lenses and screen on the optical bench with the appropriate holders. See Figure 1.

PROCEDURE, *continued*

Figure 1

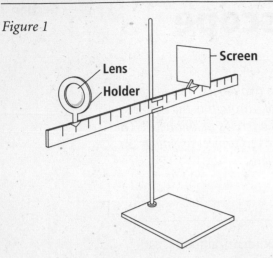

3. Point the meterstick of the optical bench at a window. Mount a lens on the end of the optical bench that is closer to the window. If the classroom does not have windows, a lamp will work. **CAUTION:** *Do not project images of the Sun.* Mount the screen on the optical bench next to the lens, on the side that is farther from the window.

4. Slide the screen along the optical bench until an image of a distant object forms. For this, you must make sure that the lenses are both precisely perpendicular to the meterstick.

5. When you have the best-focused image, read the distance between the screen and the lens. This is the focal length. Record this data for each lens.

6. Choose one lens from set A and one from set B. Mount the lenses next to each other on the window end of the optical bench.

7. Slide the lens that is farther from the window away from the other lens until you get a clear image of distant objects. Make sure the lenses are aligned perpendicular to the meterstick. When you have the best focus, the distance between the lenses is the telescope length. Record this length in Table 2.

8. You should see a relationship between the telescope length and the sum of the focal lengths of the two lenses. This is the telescope

length relation. Confirm your telescope length relation, using a different pair of lenses.

9. Choose a lens from set A, and place it at the window end of the optical bench. This is your objective lens.

10. Choose a lens from set B, and use your telescope length relation to place it on the optical bench as an eyepiece. Adjust the lens to get a sharp image of distant objects.

11. In quick succession, look at a particular distant object, first through the telescope, then directly. Estimate the ratio of the size of telescope image to the size of the image when viewed directly. This is the magnification of the telescope. Record your estimate of the magnification in Table 3.

12. Record the ratio of the focal length of the objective lens to that of the eyepiece lens.

13. Repeat steps 9–12, trying all combinations of lenses from sets A and B. You may get magnifications of 1 or even less than 1 when you use lenses from set B as objective lenses.

14. Repeat steps 9–12, using set C as objective lenses and set B as eyepieces. Record your data in Table 4.

15. Select an objective lens and eyepiece to make a telescope, using two tubes and two foam lens holders. See Figure 2.

Figure 2

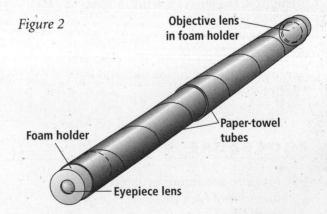

LAB 27.1

DATA AND OBSERVATIONS

Use all of the lenses from sets A (A1, A2, A3), B (B1, B2, B3), and C (C1, C2, C3) to complete the tables below.

Table 1

Lens ID	A1	A2	A3	B1	B2	B3	C1	C2	C3
Diameter (cm)									
Focal length (cm)									

Table 2

Objective ID	Focal Length of Objective	Eyepiece ID	Focal Length of Eyepiece	Sum of Focal Lengths	Telescope Length

Table 3

Objective ID	A1	A1	A1	A2	A2	A2	A3	A3	A3
Eyepiece ID	B1	B2	B3	B1	B2	B3	B1	B2	B3
Ratio of focal lengths									
Magnification									

Objective ID	B1	B1	B1	B2	B2	B2	B3	B3	B3
Eyepiece ID	A1	A2	A3	A1	A2	A3	A1	A2	A3
Ratio of focal lengths									
Magnification									

Objective ID	C1	C2	C3	C1	C2	C3	C1	C2	C3
Eyepiece ID	B1	B1	B1	B2	B2	B2	B3	B3	B3
Ratio of focal lengths									
Magnification									

LAB 27.1   **INVESTIGATION**

ANALYZE

1. Look at your data in Table 2. Compare the sum of the focal lengths to the telescope lengths. Summarize your comparison in one sentence. This is your telescope length relation.

_____

_____

2. Look at your data in Table 3. Compare the ratio of focal lengths to your estimated magnifications. Summarize this comparison in one sentence. This is your telescope magnification relation.

_____

_____

3. Use your telescope magnification relation and your measurements of focal length to predict the magnification that you would get using C2 as an objective lens with B2 as an eyepiece.

_____

4. Is your answer to question 4 greater or less than the magnification you measured using A3 as an objective lens with B2 as an eyepiece?

_____

_____

CONCLUDE AND APPLY

1. What did you notice about the effect of changing the diameter of the objective lens?

_____

_____

2. Which pair of lenses do you think would make the best telescope? Specify which lens would be the objective and which the eyepiece. Give reasons for your choice.

_____

_____

_____

_____

LAB 27.2   **DESIGN YOUR OWN**

Observing the Moon

You can estimate the phase of the Moon by observing how much of it is lit. However, a much more accurate method involves timing the precise moment when the Moon is directly south. This moment is the transit time of the Moon. Here is how it works.

For an observer in the northern hemisphere, the Sun is always in the south at noon. That is when the Sun is highest in the sky. Similarly, the southerly stars, the planets, and the Moon are all highest in the sky when they cross an imaginary line that passes through the zenith (the point straight up) and the southernmost point on the horizon. This line is called the observer's meridian. The transit time of a celestial object is when it crosses the meridian. For example, the Sun always transits near midday. When the Moon is full, it is exactly on the opposite side of Earth from the Sun. Because Earth spins once in 24 hours, the Moon transits 12 hours later than the Sun, at midnight. The quarter moon transits at 6:00 P.M. and 6:00 A.M., depending on whether it is at first quarter or third quarter. If you measure the transit time of the Moon a few times, you can predict quite accurately when the next full moon will be, or the precise length of the month.

Astronomers measure transit times with a transit telescope. They point it south, and it is free to move only up and down, perpendicular to the horizon. These telescopes have eyepieces with crosshairs so that the observer can accurately determine the transit crossing time of an object. With large objects like the Sun and Moon, the transit crossing time is when the midpoint of the object is exactly south. That moment is halfway between when the leading edge and the trailing edge of the object are exactly south.

Figure labels: Moon moving East to West · Leading edge crosses meridian · Transit · Trailing edge crosses meridian · Vertical line represents Observer's meridian · Horizontal line marks the horizon · South

PREPARATION

PROBLEM

How can you determine observationally the Moon's location at a particular time and the period of its motion?

OBJECTIVES

- **Measure** the transit time of the Moon for 1 week.
- **Predict** the times of the next full moon and quarter moon.
- **Determine** the exact length of the month.

LAB ◄ **27.2** ► **DESIGN YOUR OWN**

PREPARATION, *continued*

HYPOTHESIS

If you went to the north pole and took a spaceship
directly up, you could see Earth spinning
counterclockwise. This is why the Sun rises in the
east. Hypothesize the direction that the Moon is
orbiting Earth—clockwise or counterclockwise.
Depending on your hypothesis, will the Moon
transit earlier each day or later?

_____

_____

_____

_____

POSSIBLE MATERIALS

accurate watch
2 plumb lines (2 or 3 m) with weights
2 stepladders
simple ladder
rope
map compass with crosshairs
calculator

SAFETY PRECAUTIONS

- Under no circumstances should you look directly
  at the Sun, especially through a telescope or
  binoculars; it could cause permanent damage.
  Wear goggles to help decrease any glare.

- Do not climb on any of the ladders during this lab.

PLAN THE EXPERIMENT

As a class, design an apparatus that uses two plumb
lines to measure the transit times of the Moon.
Because the Sun transits an hour earlier on the
eastern edge of a time zone than it does on the
western edge, you cannot be sure that the Sun is
directly south at noon. For this reason, you should
keep track of the Sun's transit as well as observe the
Moon's transit. Obtaining information about the
Sun's transit from a web site will likely yield the
most current data. Do not attempt to observe the
Sun yourself. Decide how you will measure the
Moon's transits for several consecutive days. How
will you record your data? What calculations do you
need to do to find the exact transit time? You should
use 24-hour notation (hours:minutes:seconds, so
noon is 12:00:00). Outline your method below, and
draw your apparatus. Have your teacher approve
your plan before you start collecting data.

_____

_____

_____

_____

_____

LAB 27.2 **DESIGN YOUR OWN**

DATA AND OBSERVATIONS

DATA TABLES

ANALYZE

1. Did the Moon transit earlier each day or later? By how many minutes?

_____

2. Do the data support or refute your hypothesis? Which way does the Moon orbit (clockwise or counterclockwise) when viewed from above the north pole?

_____

3. Use your data to complete Table 1. Time in seconds = (hours × 3600) + (minutes × 60) + seconds. Phase in degrees = (phase in seconds) ÷ (86 400 seconds per day) × (360 degrees).

Date	Time of Sun Transit (s)	Time of Moon Transit (s)	Phase (s) (difference between times of Moon and Sun transits)	Phase (degrees)

4. Draw a circle to represent the path of the Moon's orbit. Mark a point on the circle to represent the direction of the Sun. This point is zero degrees of phase. Mark points on your circle corresponding to the phase data in degrees. These points are the positions of the Moon at the times of your measurements.

LAB <**27.2**> **DESIGN YOUR OWN**

CHECK YOUR HYPOTHESIS

Was your **hypothesis** supported by your data? Why or why not?

_____

_____

CONCLUDE AND APPLY

1. How many seconds passed between the first transit of the Moon you observed and the last one?

_____

2. Through how many degrees did the Moon move between the first and last timings of the Moon?

_____

3. Using the answers to questions 1 and 2, how many seconds does it take the Moon to move through one degree?

_____

4. How much time would it take the Moon to move through 360°? Express your answer in seconds. Then express your answer in days, hours, minutes, and seconds. This calculation shows the length of a month. Does it make sense?

_____

_____

5. When will the next full moon be? Compare your prediction with the time stated in a calendar or on a web site. What about the next quarter moon and new moon? How good are your predictions?

_____

_____

_____

LAB **28.1**   **INVESTIGATION**

Your Age and Weight on Other Planets

What could be more down-to-earth than a person's age and weight? Yet both are controlled by astronomical forces. The Sun swings Earth around once, and we say that another year has gone by. But how many birthdays would you have had if you had been living on Mercury? The pull between Earth and a human body leads us to declare that a person weighs some amount. How heavy would you be if you lived on Venus? You can use Kepler's third law and Newton's law of universal gravitation to answer these questions.

PREPARATION

PROBLEM
What would your age and weight be if you lived on another planet?

OBJECTIVES
- **Calculate** your age on the other eight planets of the solar system.
- **Calculate** your weight on each planet.

MATERIALS
calculator
scale

PROCEDURE

1. Kepler's third law states that the orbital radius (a) of a solar system planet relates to its orbital period (P) in this formula if P and a are expressed in solar system units: $P^2 = a^3$. In solar system units, the unit of time is the year, the unit of length is the astronomical unit (AU), and the unit of mass is the mass of the Sun. Use this relation and the data in Table 1 to find out the length of the planet's year in Earth years.

2. You now have the number of Earth years per planet year, but what you want is the number of planet years per Earth year. So calculate the reciprocal.

3. Multiply the number of planet years per Earth year by your age to obtain your age on the planet. Record your results, to three decimal places, in Table 2.

LAB **28.1**   **INVESTIGATION**

PROCEDURE, *continued*

4. To calculate your weight on another planet, use Newton's law of universal gravitation.

$$F = G\frac{m_1 m_2}{r^2}$$

F is the force between two bodies, which in this case is your weight on the planet; G is the universal constant of gravitation; m_1 is the mass of the planet; m_2 is your own mass; and r is the distance between the centers of the two bodies, which in this case is the radius of the planet. Notice that even though your weight is different from planet to planet, your mass remains the same.

5. Using the scale, weigh yourself to find your weight in pounds. Convert your weight from pounds (lb) to kilograms (kg). Use the following formula to make your calculations. 1 lb = .455 kg

6. Using Table 1, and Newton's law of universal gravitation in step 4, calculate your weight on each planet. m_1 is your mass, which you calculated in step 5. m_2 is the mass of the planet. $G = 6.6726 \times 10^{-11}$ m³/kg•s².

7. Record your results in Table 2.

LAB 28.1   **INVESTIGATION**

DATA AND OBSERVATIONS

Table 1

Planet	Orbital Radius, a (AU)	Planetary Radius, r (km)	Planetary Mass, m (10^{24} kg)
Mercury	0.387	2439.7	0.3302
Venus	0.723	6051.8	4.8685
Earth	1.0	6378.1	5.9736
Mars	1.524	3397	0.64185
Jupiter	5.204	71 492	1898.6
Saturn	9.582	60 268	568.46
Uranus	19.201	25 559	86.832
Neptune	30.047	24 764	102.43
Pluto	39.236	1195	0.0125

Table 2

Planet	Your Age in Planet Years	Planet Mass/ Earth Mass (kg)	Earth Radius/ Planet Radius (km)	Square of Radius Ratio (km)	Your Weight on Planet (N)
Mercury					
Venus					
Earth					
Mars					
Jupiter					
Saturn					
Uranus					
Neptune					
Pluto					

LAB **28.1**   **INVESTIGATION**

ANALYZE

1. In the years of which planets are you oldest and youngest? Is this surprising? Why or why not?

_____

_____

_____

2. On which planet would you be heaviest?

_____

_____

CONCLUDE AND APPLY

1. What is the weight of the heaviest thing you can lift? If an object weighed that much on the planet on which you are heaviest, how much would it weigh on Earth? Name an object that weighs this much.

_____

_____

_____

_____

2. What would be the heaviest object you could lift on Pluto?

_____

_____

_____

Use with
Section 28.3

Relating Gravitational Force and Orbits

*I*n the seventeenth century, Kepler noticed that the planets' periods of revolution
are related to their orbital distances in a special way. His observation later became
known as his third law: $P^2 = a^3$. Would this relation be true if the Sun were twice as
massive as it is? Or half as massive? Does the orbital period depend, not only on the
distance at which the planet orbits, but on the mass of the central body? We can use
a model to investigate these questions.

PREPARATION

PROBLEM

How do the orbital periods of the
planets relate to the mass of the Sun?

OBJECTIVES

- **Construct** a simple model of planetary motion.
- **Check** the model to see how well it
  follows Kepler's third law.
- **Use** the model to find out if orbital
  period depends on the mass of the
  central body.
- **Estimate** the mass of Earth.

HYPOTHESIS

With your group, form a hypothesis
about whether changes in the central
mass affect Kepler's third law.

_____

_____

_____

_____

POSSIBLE MATERIALS

glass tube, 15 cm, fire polished
   and taped
scissors
duct tape
fishing line
plastic-foam ball, 10 cm
paper clips
40 metal washers
metric ruler
stopwatch
marking pen
calculator

SAFETY PRECAUTIONS

Be careful when handling scissors.
Never point sharp objects at anyone.

 **LAB 28.2** **DESIGN YOUR OWN**

PLAN THE EXPERIMENT

Design an experiment to study the effect of varying the central mass, using a small-scale model. The figure below shows a suggestion for a model. How will you show whether the model follows Kepler's third law? What will your independent and dependent variables be? What measurements and data do you need to gather? How will you test the effect of different mass on the relationship? Outline your experiment. Design a table to record your data. Have your teacher approve your plan before you start the experiment.

_____

_____

_____

_____

_____

_____

_____

_____

_____

_____

_____

_____

_____

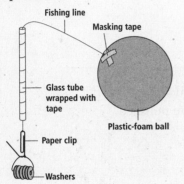

Fishing line

Masking tape

Glass tube wrapped with tape

Plastic-foam ball

Paper clip

Washers

DATA AND OBSERVATIONS

Sample Data Table

Radius (cm)	Swings	Total time (s)	Period (s)	p^2/a^3 ($\times 10^{-15}$)

LAB 28.2   **DESIGN YOUR OWN**

DATA AND OBSERVATIONS, *continued*

DATA TABLES

LAB **28.2**   **DESIGN YOUR OWN**

ANALYZE

1. Does the period increase, decrease, or stay the same as you increase the radius of revolution with the central mass fixed?

_____

2. With the radius fixed, does the period increase, decrease, or stay the same as you increase the number of weights?

_____

_____

3. Did your results meet your expectations? Explain your answer.

_____

_____

CHECK YOUR HYPOTHESIS

Was your **hypothesis** supported by your data? Why or why not?

_____

_____

CONCLUDE AND APPLY

1. How well do you think your model performed? Would you draw strong conclusions from this model? Explain your answer.

_____

_____

_____

_____

2. Assuming *a* is constant, what happens to the square of the period (P^2) when you double the number of weights? Does it triple, double, stay the same, half, or third?

_____

_____

3. Assuming *a* is constant, how long do you think an Earth year would last if the Sun were twice as massive as it is?

_____

_____

LAB ◀ **29.1** ▶   **INVESTIGATION**

Diameter and Rotation of the Sun

Although the Sun is a fairly typical star, it is very special to us because it is much closer to Earth than any other star. That makes it the easiest star to study. With simple equipment, you can measure the diameter and rotation of the Sun.

PREPARATION

PROBLEM
What are the diameter and rotation rate of the Sun?

OBJECTIVES
- **Measure** the diameter of the Sun.
- **Measure** the rotational rate of the Sun.
- **Estimate** the size of sunspots.

MATERIALS
meterstick
2 index cards
scissors
tape
aluminum foil, 4 cm square
straight pin

single-edge razor blade
unlined white paper
clipboard
small telescope
small telescope stand

SAFETY PRECAUTIONS

- Wear safety goggles during the lab procedure to help eliminate glare.
- Never look at the Sun through a telescope or directly at the Sun; it could blind you.
- Use caution when handling pins, scissors, or razor blades. Be careful not to puncture or cut your skin.

PROCEDURE

1. With a razor blade, make three slits in the shape of a capital I in two index cards so that they can be mounted snugly on a meterstick.

2. Draw a pair of fine parallel lines on one of the index cards, exactly 0.8 cm apart. You will project an image of the Sun onto this screen.

3. Cut a 1-cm × 1-cm-square hole in the other index card and tape aluminum foil over it. Use a pin to punch a small, clean hole in the center of the aluminum foil.

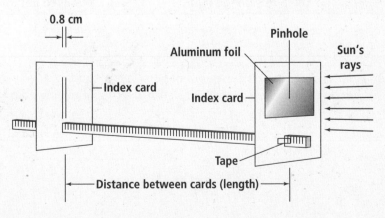

Figure 1

LAB 29.1 **INVESTIGATION**

PROCEDURE, *continued*

4. Assemble the pinhole camera as shown in Figure 1. Make sure that the cards fit tightly on the meterstick and that they are perpendicular to it.

5. Point the pinhole at the Sun so that an image of the Sun projects onto the screen.

6. Move the screen along the meterstick until the circular image of the Sun has a diameter of exactly 0.8 cm.

7. Measure the distance between the pinhole and the image. If the cards are perpendicular to the meterstick, this will be the distance between the cards. Record this distance and the diameter of the Sun's image in the table provided.

8. Fit the telescope with a low-power eyepiece if possible. Make sure the lens cap is on it.

9. With the lens cap *on* the telescope, mount the telescope on its stand. Point the telescope at the Sun. In order to avoid looking directly at the Sun, the telescope's shadow can be used to fine-tune the pointing.

10. Make a screen by putting a piece of unlined paper on a clipboard. Place the screen next to the eyepiece and carefully remove the lens cap.

Adjust the alignment until the projection is visible.

11. Move the screen away from the eyepiece until the image of the Sun is almost as wide as the paper. You may need to adjust the alignment and focus of the telescope for this. The screen may be several meters from the telescope at this point, depending on the power of the telescope.

12. Focus the telescope until you have a sharp image of the Sun. You should see clusters of sunspots. Trace the outline of the Sun directly onto the screen paper. Trace the sunspots, making a careful record of their locations and sizes. The image will tend to drift across the screen because Earth is rotating.

13. Draw an arrow to show which direction the image is drifting. Record on the drawing the date and time. Write down any interesting observations.

14. Repeat steps 8–13 at the same time of day for 4 or 5 days, drawing on separate sheets of paper. Try to make the size of the Sun's image the same in each drawing.

DATA AND OBSERVATIONS

Table

Diameter of the Sun's Image (cm)	Distance Between Pinhole and Screen (cm)

LAB **29.1**   **INVESTIGATION**

ANALYZE

1. Figure 2 has two similar triangles. Similar geometric shapes have proportional sides, so the diameter of the Sun (km)/distance to the Sun (km) = diameter of the Sun's image (cm)/distance from the pinhole to the screen (cm). The distance to the Sun is about 1.5×10^8 km. Calculate the diameter of the Sun.

_____

_____

Figure 2

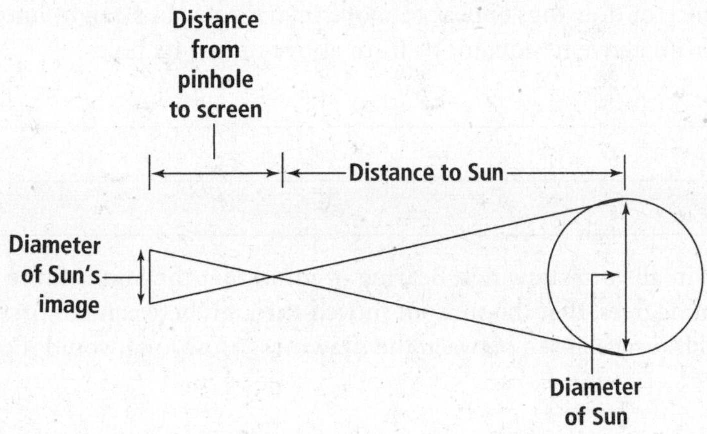

Distance
from
pinhole
to screen

◄—— Distance to Sun ——►

Diameter
of Sun's
image

Diameter
of Sun

2. Do you think the orientation of the Sun is the same in each of your drawings? Give reasons for your answer.

_____

_____

_____

3. Choose an individual sunspot near the center of one of your drawings. What is its diameter, in centimeters? What is the diameter of the Sun, in centimeters? What is the ratio of the diameter of the sunspot to that of the Sun?

_____

_____

_____

4. Multiply the ratio of the diameter of the sunspot to that of the Sun by the actual diameter of the Sun to determine the actual size of the sunspot, in kilometers. Compare this to the size of Earth.

_____

_____

_____

LAB **29.1**   **INVESTIGATION**

CONCLUDE AND APPLY

1. Would sunspots appear to move in circles or along straight lines if we were viewing the Sun from directly above its equator? How about if we were viewing from directly above one of the Sun's poles?

_____

_____

2. Do the sunspots in your sequence of drawings appear to move in circles or in straight lines? Do you think we are viewing the Sun from above its equator or from above one of its poles?

_____

_____

_____

3. Choose a sunspot that appears in all your drawings. Bearing in mind that the Sun is more or less spherical, estimate the angle, in degrees, that the sunspot moved through between the first drawing and the last drawing. How much time elapsed between the drawings? How long would it take for the sunspot to move through 360°?

_____

_____

_____

LAB ◄ **29.2** ►   **MAPPING**

Use with
Section 29.2

Constellations and the Seasons

Even in a city, where lights make many stars invisible, you can see a few hundred stars in various constellations at night. Star charts identify many well-known stars and constellations, even as they move across the sky in an evening and from season to season.

PREPARATION

PROBLEM
How do stars appear to move in the sky?

OBJECTIVES
- **Identify** several stars and constellations in the night sky.
- **Understand** how stars move during a night.
- **Understand** why different constellations are visible during a year.
- **Measure** the latitude of your city or town.

MATERIALS
protractor with hole at origin
stiff, thin wire, 10 cm
small weight
tracing paper
binoculars

SAFETY PRECAUTIONS

- Use caution when handling wire. Be careful not to puncture or cut your skin.
- Wear safety goggles during the lab procedure.

PROCEDURE

1. Make a plumb line by attaching a weight to a wire.

2. Thread the free end of the wire through the hole at the origin of a protractor. Bend the wire over. If you hold the protractor with the flat edge at the top, the plumb line should hang free, allowing you to read angles on the protractor scale.

3. Take your protractor and plumb line outside shortly after sunset. Use the star charts in this lab to find the pointers of the Big Dipper (Ursa Major). These help you locate Polaris (the North Star).

4. Sight along the flat edge of the protractor toward Polaris. Read the altitude of Polaris (the angle of Polaris above the horizon) from

the protractor scale. Record in Table 1 the altitude of Polaris, and note a feature on the horizon that is below Polaris.

5. A position angle is the angle that the line joining two points makes with the vertical. For example, the position angle of a clock's hour hand at three o'clock is 90°. At eight o'clock, the position angle is 240°. Use the protractor with the scale facing you. Measure the position angle of the pointers in the Big Dipper. Record this angle and the time.

6. Repeat steps 3–5 after 2 hours, after 4 hours, and after 6 hours if it is practical.

LAB ◁ **29.2** ▷   **MAPPING**

PROCEDURE, *continued*

7. On the next day in class, answer Analyze questions 1–4.

8. Use the star charts to identify the compass direction of each constellation in Table 2 in each season.

9. Using tracing paper, draw the circular outline of the autumn map and mark the compass directions on it. Label this "Map 1: Seasonal Motion of Gemini." Trace the constellation Gemini on the map and label it "Autumn." Place the tracing paper on the winter map, lining it up with the compass directions, and trace Gemini again, labeling it "Winter." Repeat for "Spring." (Note that Gemini does not appear on the summer map.)

10. Draw arrows between the seasonal locations of Gemini. Answer Analyze question 5.

11. The three stars Vega, Deneb, and Altair constitute the summer triangle. Label a piece of tracing paper "Map 2: Seasonal Motion of Summer Triangle." Use the method in step 9 to trace the seasonal locations of the summer triangle.

12. Label a piece of tracing paper "Map 3: Seasonal Motion Around Polaris." Using the spring map, trace the positions of Polaris and Cassiopeia. Label them "Position 1." Repeat this, using the winter map and labeling them "Position 2." Draw an arrow showing the direction of motion from Position 1 to Position 2.

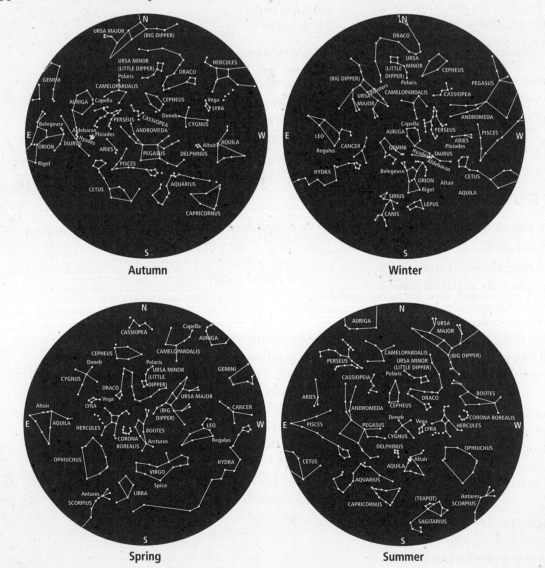

Autumn

Winter

Spring

Summer

LAB **29.2** **MAPPING**

DATA AND OBSERVATIONS

Table 1

Time	Altitude of Polaris	Horizon Feature Below Polaris	Position Angle of Pointers

Table 2

Constellation	Autumn	Winter	Spring	Summer
Bootes				
Lyra				
Orion				
Pegasus				

ANALYZE

1. What is remarkable about the position of Polaris?

_____

2. Does the Big Dipper move clockwise or counterclockwise around Polaris?

_____

_____

3. Use the elapsed time between your first and last measurement of the position angle of the Pointers and the angle that the Pointers moved through during that time to calculate how much time it takes for the Pointers to move through 1°.

_____

4. How much time does it take for the Pointers to move through 360°? What is the significance of this answer?

_____

_____

_____

5. In what direction does Gemini move throughout the seasons?

_____

LAB **29.2** MAPPING

CONCLUDE AND APPLY

1. Use a local map to find out the latitude from which you observed Polaris. Compare the latitude to your measurement of the altitude of Polaris. What do you conclude about the relationship between the latitude of the observer and the altitude of Polaris?

   _____
   _____
   _____
   _____

2. Based on the path of Gemini in Map 1, where would Gemini be in summer? Explain why Gemini does not appear on the summer map.

   _____
   _____
   _____
   _____
   _____

3. In what season does the summer triangle appear in view? In what direction does it move across the sky? In what season does it disappear from view?

   _____
   _____
   _____

4. Map 3 shows the positions of Cassiopeia at two times of the year. Cassiopeia also passes through these positions at different times of day. Suppose Cassiopeia was at Position 1 at 6:00 P.M. At what time would it be at Position 2?

   _____

LAB ◀ **30.1** ▶   **INVESTIGATION**

Modeling Spiral Galaxies

Use with
Section 30.1

Galaxies come in all shapes and sizes. Some, like the Milky Way Galaxy, have a broken spiral-arm pattern. Astronomers theorize that the arms in this type of galaxy continually form and re-form in response to supernovae explosions and other disturbances in space. Other galaxies have distinctive two-arm spiral patterns that appear to be permanent features. Scientists hypothesize that these arms may be maintained by spiral density waves that rotate through interstellar gas and dust in rigid patterns.

Barred spiral galaxy

PREPARATION

Normal spiral galaxy

PROBLEM

How do the arms of spiral galaxies form?

OBJECTIVES

- **Model** spiral galaxies.
- **Compare** and **contrast** scientific theories about the formation of spiral arms.
- **Describe** the characteristics of spiral galaxies.

MATERIALS

water
oil
teaspoon
nondairy powdered creamer
250-mL beaker
several rocks
bucket

SAFETY PRECAUTIONS

Wear safety goggles and an apron during the lab.

PROCEDURE

1. Working with a partner, pour 200 mL of water at room temperature into the beaker.

2. Add a teaspoon of oil to the water. Then carefully add a teaspoon of powdered creamer to the mixture.

3. Use the teaspoon to slowly stir the mixture in a clockwise direction. Observe what happens to the creamer. While you are stirring, your partner should sketch your observations in the box labeled Sketch 1 in Data and Observations.

4. Stop stirring the mixture. Observe what happens to the creamer. Sketch your observations in the box labeled Sketch 2 in Data and Observations.

5. Next, fill a bucket with water. Drop a rock into the water, then sketch your observations in the box labeled Sketch 3 in Data and Observations.

LAB 30.1 **INVESTIGATION**

DATA AND OBSERVATIONS

Sketch 1

Sketch 2

Sketch 3

LAB 30.1 INVESTIGATION

ANALYZE

1. Describe your observations of the mixture. What happened to the creamer when you stirred the mixture? What happened when you stopped stirring?

   _____
   _____
   _____
   _____

2. What did you observe when the rock was dropped into the bucket of water? What astronomical event did the rock model? Explain your answer.

   _____
   _____
   _____
   _____

3. Explain how your observations after stirring the mixture resemble a spiral galaxy. How do your observations differ from a spiral galaxy?

   _____
   _____
   _____
   _____

4. Identify the parts of a spiral arm galaxy and state where each part is located. Where would globular clusters be found?

   _____
   _____
   _____
   _____

LAB **30.1** INVESTIGATION

CONCLUDE AND APPLY

1. Which theory of spiral arm formation does your observations from the procedure support? Explain.

_____

_____

_____

_____

2. Use your observations of the rock and the bucket of water to form a hypothesis about what would happen to interstellar dust and gas following a supernovae.

_____

_____

_____

3. Where would young stars be located in your model galaxy? Where would old stars be?

_____

_____

_____

4. An irregular galaxy has no distinct shape. How could you model an irregular galaxy?

_____

_____

_____

LAB 30.2 MAPPING

Three-Dimensional Map of the Local Group

The Milky Way Galaxy is just one of a number of galaxies in our cosmic neighborhood, which are collectively known as the Local Group. The galaxies of the Local Group are believed to be gravitationally bound together, so they perform a complicated orbit around one another. By far, the biggest two galaxies in the Local Group are the Milky Way and M31, the Andromeda Galaxy, which is about 2 million ly away.

The locations of the galaxies of the Local Group are in the table in Data and Observations. The x, y, and z coordinates are given for each galaxy in units of 100 000 ly. Earth is at the origin of the coordinate system. The z-axis points north. The x-axis points in the direction of the Sun, as viewed from Earth on the vernal equinox (March 21).

PREPARATION

PROBLEM

What does the Local Group of galaxies look like?

OBJECTIVES

- **Map** the Local Group from three viewpoints.
- **Construct** a scale model showing the locations of the galaxies of the Local Group.

MATERIALS

cardboard, 30 cm × 20 cm
fishing line or thread, 4 m
metric ruler
scissors
heavy needle
modeling clay

SAFETY PRECAUTIONS

Be careful when using sharp objects to make holes in the cardboard. Pointed objects can puncture skin.

PROCEDURE

1. Look at the three blank maps in Analyze. Each map will represent a view of the Local Group along one of the axes. On the map labeled "*xy* projection," plot the position of the Milky Way Galaxy and M31, using the *x* and *y* coordinates in Table 1.

2. Plot the positions of the Milky Way and M31 on the *yz* projection map using the values for *y* and *z* for each galaxy and on the *xz* projection map using the values for *x* and *z* for each galaxy. Have your teacher check your graphs.

3. Mark the Milky Way and M31 Galaxies in a bright color. Plot the locations of all the other galaxies on all three projections.

LAB 30.2   **MAPPING**

PROCEDURE, *continued*

4. Draw a scale copy of the *xy* projection on a piece of cardboard. Use a scale of 0.3 cm/ 100 000 ly. Cross-check your copy with your original map of the *xy* projection. When you are satisfied that the locations are correct, push a hole through each point with a needle.

5. The cardboard sheet will be the top of your model of the Local Group. You will suspend the galaxies on threads from the horizontal cardboard sheet. For each galaxy, the length of its suspending thread should be $(40 - z) \times 0.3$ cm, where z is the coordinate of the galaxy

in units of 100 000 ly. This assumes that you are using a scale of 0.3 cm/100 000 ly. Cut each thread a little longer than this length. Make a knot at one end of the thread and pass the thread through the appropriate hole in the cardboard sheet. Trim the thread to the correct length and fix a clay "galaxy" to the lower end of the thread.

6. When all the "galaxies" are in place, make four extra holes in the cardboard, one in each corner, and use these to suspend the model with four pieces of thread.

DATA AND OBSERVATIONS

Galaxy Name	Distance in 100 000 ly		
	x	y	z
WLM	19.27	0.17	−5.33
IC 10	20.33	1.81	34.4
NGC 147	14.42	2.1	16.48
Andromeda III	17.47	2.71	13.09
NGC 185	14.42	2.47	16.43
NGC 205	16.18	2.87	14.63
M32	16.35	3.08	14.39
M31	16.25	3.06	14.51
Andromeda I	16.99	3.43	13.54
SMC	0.86	0.20	−2.87
Sculptor	1.61	0.43	−1.11
Pisces	26.74	7.63	11.26
IC 1613	23.99	6.98	0.92
Andromeda II	17.35	6	12.12
M33	19.73	8.57	12.74
Fornax	3.16	2.64	−2.83

Galaxy Name	Distance in 100 000 ly		
	x	y	z
LMC	0.11	0.68	−1.88
Leo A	−37.16	21.58	25.56
Carina	−0.34	1.86	−2.33
Leo I	−5.18	2.74	1.28
Sextans I	−2.68	1.35	−0.09
Leo II	−5.44	1.12	2.26
GR8	−37.5	−9.91	9.78
Ursa Minor	−0.79	−0.86	2.76
Draco	−0.28	−1.57	2.54
Milky Way	−0.02	−0.26	−0.15
SagDIG	14.58	−35.21	−12.15
NGC 6822	7.26	−14.75	−4.33
DDO 210	19.48	−21.82	−6.67
IC 5152	10.91	−6.12	−15.6
Pegasus	47.90	−6.60	12.74

LAB 30.2   **MAPPING**

ANALYZE

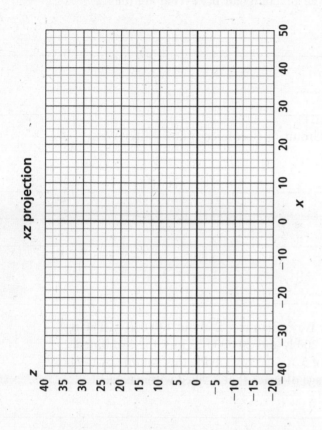

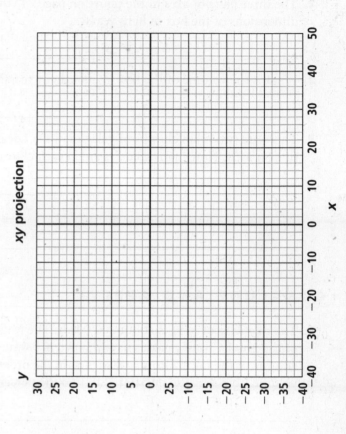

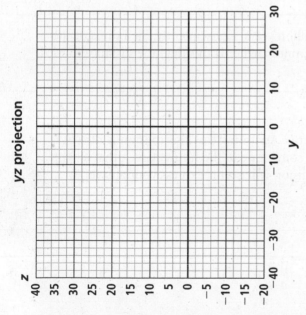

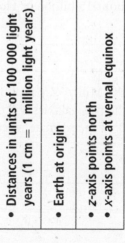

- Distances in units of 100 000 light years (1 cm = 1 million light years)
- Earth at origin
- z-axis points north
- x-axis points at vernal equinox

LAB ⟨ **30.2** ⟩ **MAPPING**

ANALYZE, continued

1. The three pairs of axes in the figure on page 247 define a rectangular box. What are the dimensions of the box in light years?

_____

2. What is the volume of the box in cubic light years?

_____

3. There are two main clumps of galaxies in the Local Group, one centered on the Milky Way, one on M31. How many galaxies are in each clump?

_____

_____

_____

CONCLUDE AND APPLY

1. The diameter of M31 is about 200 000 ly. The Milky Way's diameter is about 130 000 ly. The rest of the Local Group galaxies have diameters between 200 ly and 20 000 ly. Thus, the combined volume of the Local Group galaxies is on the order of 5×10^{15} ly^3. Divide this volume by the box volume from Analyze question 2 to find the percentage of space that is filled with galaxies.

_____

_____

2. The age of the universe is estimated to be about 15 billion years. Consequently, the radius of the observable universe is about 15 billion ly. How many times farther is this than the distance to M31?

_____

_____

3. M31 is moving toward the Milky Way at about 100 km/s. Assuming that this motion is *directly* toward the Milky Way, the two galaxies will collide at some time in the future. Given that a light year is equal to about 10^{13} km, estimate when the collision will occur. Express your answer in seconds and in years.

_____

_____

Credits

ART CREDITS
Glencoe: **x, xi, xii**; Navta Associates: **1, 2, 33, 45, 51, 63, 64, 67, 75, 83, 190, 191, 193, 206, 247, 266, 268, 270, 272, 283, 291, 313**; MacArt Design: **12, 23, 38, 39, 51, 53, 54, 57, 58, 60, 61, 65, 66, 67, 73, 97, 105, 109, 114, 122, 125, 129, 133, 158, 165, 181, 182, 183, 186, 188, 197, 206, 218, 221, 230, 233, 235, 238, T282, T292, T295, T296, T297, T299, T302**; MapQuest.com: **16**; Mapping Specialists: **146, 154, 155, T288, T289**

PHOTO CREDITS
233 NASA